The Climate Misinformation Crisis

How to Move Past the Mistruths to a Smarter Energy Future

Tushar Choudhary, Ph.D.

HopeSpring Press

Published by HopeSpring Press

Rosenberg, Texas

Library of Congress Control Number: 2023923515

Paperback ISBN: 979-8-9864358-3-1

eBook ISBN: 979-8-9864358-2-4

Cover Design Credit: Rajeshree Choudhary

To the many who strive to keep an open mind.

Acknowledgements

Several people have made many valuable contributions to this book. I would like to take the opportunity to thank them.

The idea to write this book originated from discussions with my dad, Dr. Vasant Choudhary. I would like to thank him for his encouragement. I would also like to thank him and several colleagues for their technical input.

I would like to thank Bhupinder Singh, Dr. Kiran Rao, and Archana Ghonasgi for their insightful suggestions. The book is more reader friendly because of their help.

Many thanks to my wife, Rajeshree Choudhary. She has contributed to the cover design and formatting of the book. I could not have written this book without her constant support and encouragement.

Finally, I would like to thank my sister, Rashmi Sarode, for proof reading.

...§§§-§-§§§...

Table of Contents

Preface

Conversations with my dad[1] are not typical. Our discussions are mainly about issues related to energy. Perhaps this is because we have dedicated a combined seventy years to energy R&D.

Energy solutions for climate change have been the focus of the discussions in recent years. These discussions invariably reach the same conclusions:

- Extreme views from both sides are causing a misinformation crisis.
- Sustainable progress is not likely until the impact from the misinformation is mitigated.

I have written this book to discuss a) why misinformation is so dangerous, and b) how to make progress in this critical area.

The balanced discussions might be uncomfortable to those with extreme views. That is not my intent, but balanced discussions are essential to meet the goals of the book.

I believe that my experience and passion have equipped me well to provide a balanced view. My experience includes projects related to both low carbon energy and fossil fuel energy. Since retiring from the energy industry in early 2019, I have declined offers to consult for the fossil fuel industry as well as the green industry. I am not financially obligated to any special interests or research grants. I donate the earnings from my books to charity. My passion is to enable the best energy solutions to meet global challenges. These attributes allow a bias-free environment for my assessments.

This work has been made easier by my earlier books: a) Critical Comparison of Low-Carbon Technologies (2020) and b) Climate & Energy Decoded (2022).

Tushar Choudhary, Ph.D.
Houston, Texas

...§§§-§-§§§...

Introduction

War: Struggle, hostility, or competition between opposing groups for a particular end[2].

Humans have fought many wars over the ages. These wars have been fought over resources, religion, politics, or ideology.

Currently, we are amidst an ideological war that has massive global implications. This war is being waged by groups with opposing views about climate and energy.

What are the beliefs and goals of the two groups? How does the public fit in this war? What needs to be done to ensure the best possible future?

I will attempt to answer these questions in this book.

The Climate & Energy War

The ideological war is being fought over two key questions.

Are human activities causing a serious climate change problem[3]? And how to meet the growing energy demand of the global population?

Climate activists and climate skeptics are the opposing groups in this war. I will refer to them as activists and skeptics[4]. Activists and skeptics have opposing views on climate and energy.

Activists believe that climate change–caused by human activities such as fossil fuel use–is our most serious problem[5]. They want an immediate, drastic shift from fossil fuels to low carbon energy.

In contrast, skeptics believe that climate change is not a serious problem. They believe that the benefits from fossil fuels far outweigh the negative impacts. So, they want to maintain the status quo.

The term "serious problem" can have a different meaning to different people. I will take a moment to discuss my definition. A serious problem is something that is very important to the well-being of global society and requires urgent attention. For example, the lack of safe water and clean cooking fuels for billions, and erosion of confidence in elections are serious problems by my definition.

A major goal for both groups–activists and skeptics–is to gain support from the public. Such support can push energy policies that can enable their desired goals.

An old proverb comes to mind[6]. "*The first casualty of war is truth*". This proverb is especially relevant to the climate and energy war. Both warring groups are liberally using misinformation to gain support of the public. In fact, the primary weapon in the war is misinformation.

This war is being fiercely fought in mainstream media and in social media. It has created a misinformation crisis. This famous quote is sobering[7]. "*A lie can travel halfway around the world while the truth is putting on its shoes*".

How we deal with this crisis will decide the future of energy and the environment. Fortunately, the public has the power to decide the outcome.

For a good outcome, the public must avoid the misinformation trap. The global society must move past the misinformation crisis. This will enable the best possible future for our planet.

Key Elements of the Book

My goal in this book is to share the most credible data available to address the mistruths about climate and energy[8].

The book will provide information that will help the readers to weed out misinformation. To accomplish this, I will provide crucial facts about climate change, fossil fuels, renewable energy, and the low carbon transition. I will also specify the principles that can be used to avoid misinformation.

Finally, I will discuss how a rational approach to climate and energy can ensure a brighter future. This discussion will focus on the importance of focusing on **both science and realities of climate and energy**. *Realities* are practical aspects that cannot be ignored. Consider our ordeal with the COVID-19 virus when infections were at their peak. Based on science, the most efficient way to beat the disease would have been to isolate everyone for several weeks. This would have prevented the transfer of the disease and stopped it in its tracks. The reality was that complete isolation would have caused chaos in the society. Essential workers had to be exempt from isolation to provide medical, energy, water, security, and food services. This was a reality that could not be ignored. The science as well as the realities had to be considered. Similarly, it is crucial to be consistent with science and realities to address any complex problem.

For this book, activists and skeptics are those whose views are not consistent with the science or realities of climate and energy[9].

...§§§-§-§§§...

1. Basics of Misinformation

Misinformation includes false and misleading information. An entity that spreads misinformation may not have harmful intent. The entity may not know that the information is false or misleading. But that does not make it less dangerous.

Let us consider an example.

Jane informs a friend, who is a new mother, that her child should not get a polio vaccine because it is not safe. Jane believes that the information is accurate based on social media posts. She means no harm to the mother or her child. In fact, she is trying to help. Nevertheless, her false information can destroy the life of the child[10].

We need to avoid misinformation to make robust decisions. To avoid misinformation, we must first recognize it.

False information is often misaligned with advertised facts or logic. So, it is easier to recognize false information.

In contrast, misleading information is more difficult to recognize. This is because it can appear to be factual and logical. This makes it more convincing compared to false information. Misleading information can be more dangerous because of this reason.

How to circumvent this challenge? The first step is to recognize the different types of misleading information.

I will discuss the major types in this chapter.

Misleading by Key Omissions

This is the most common type. In this type, partial or selective data is used to discuss a topic. Specifically, key data that is essential for a robust discussion is omitted. Such exclusions lead to misleading conclusions.

Below are four common issues:

- Only the positive aspects are discussed. Benefits are advertised, but the associated problems are not discussed. Such misleading information is used to promote an idea or a product.
- Only the negative aspects of an idea or product are discussed. This is used to mislead about the competition or promote hatred.

- Only the worst-case scenario is discussed when discussing the future. Realistic scenarios are ignored. This is used to promote fear or sensational news.
- Only the best-case scenario is discussed when discussing the future. Realistic scenarios are ignored. This is used to hype solutions and promote unrealistic optimism.

Misleading by Biased Assumptions

In this type, biased assumptions are used in the calculations.

Complex calculations require many assumptions. Bias is introduced via the use of improper assumptions. The calculations are sensitive to the assumptions. Use of biased assumptions leads to estimates that are not realistic. The choice of assumptions determines the results of the estimates. Thus, a desired result can be obtained by choosing specific assumptions.

Let us look at an example.

Jack and Jill are an average couple in their early 50s. The couple has modest retirement savings which they have invested in the stock market. Jill believes that their retirement savings are not adequate. She wants to cut their spending to increase their retirement savings. Jack is not keen on cutting expenses. He develops a detailed calculation to assist with the decision. In his calculations, he uses the following assumptions: a) an unusually large increase in his salary every year until he retires, b) top quartile performance of the stock market over the next five decades and c) very low medical expenses in retirement. His estimate suggests that a spending cut is not required.

Jack used overly optimistic assumptions for his estimates. These biased assumptions were designed to support his narrative. What if he had used realistic assumptions–such as a reasonable salary increase, average stock market performance, and moderate medical expenses? If so, the results would have been markedly different. The results would have shown the need for a large cut in spending. This example shows how biased assumptions can lead to poor decisions.

Misleading by Emphasis on Outliers

In this type, undue importance is given to outliers.

Outliers refer to unique situations or special circumstances. The information that is based on outliers is not broadly applicable.

When such information is hyped without highlighting the outlier aspects, it gives the impression that the information is broadly applicable.

Let us consider an example.

Tom does very little physical activity and has poor food habits. He routinely uses the former U.S. President Donald Trump as an example to hype that physical activity and a good diet are not necessary for good health[11]. But Mr. Trump is an outlier. Tom ignores the medical consensus that most people need to be physically active and have good food habits to remain healthy[12,13]. Instead, he places emphasis on outlier data.

Understanding the basics of misinformation equips us to ask the right questions. These questions–which will be discussed in a later chapter–help to identify and avoid misinformation.

...§§§-§-§§§...

2. Setting the Stage

Here, we will review the key disputes between activists and skeptics. This chapter will also focus on why activists and skeptics are so heavily involved with misinformation.

Climate Change

Climate change is defined as a long-term shift in temperature and weather patterns[14].

Natural causes have been changing our climate since the beginning of time[15,16]. Examples of natural causes include solar variations, orbital changes, volcanic activity, and changes in earth's reflectivity[17].

The dispute is about the climate change caused by human activities. The main human activity of concern is energy production from fossil fuels. This has led to very large emissions of greenhouse gases over the last few centuries.

Henceforth, I will refer to climate change caused by natural causes as "natural climate change" and *climate change caused by human activities as "climate change"*.

Activists believe that climate change is the most serious problem facing humanity. Many believe we are at the threshold of extreme devastation because of climate change.

Skeptics do not believe that climate change is a serious problem. They believe that we are mainly experiencing natural climate change.

Energy

Currently, fossil fuels meet over 80% of the global energy demand[18]. The massive use of fossil fuels leads to very large emissions of greenhouse gases. A shift to low-carbon energy can drastically reduce greenhouse gases. Such a shift will require at least two steps.

- Fossil fuel power will need to be replaced with low carbon power. Examples of low carbon power are solar, wind, nuclear, hydro and bio power.

- Technologies that use fossil fuels will need to be replaced with technologies that use electricity. This is known as electrification. Examples of technologies that use electricity are electric vehicles and heat pumps.

Activists are pushing for an immediate shift from fossil fuels because of their concern about climate change. They want a very rapid shift to low carbon power. They also want to focus on very rapid electrification at the same time. In their view, such a shift will be straightforward. They believe that the main challenge is the lack of human will.

Skeptics do not want to move away from fossil fuels because they do not believe that climate change is a serious problem. Also, skeptics have a poor view of low carbon technologies. In their view, a shift to low carbon energy can cause great harm to society.

Why So Much Misinformation?

Why do the two groups have opposing views? Why is there no willingness to listen to each other? Why do they focus only on the information that supports their narratives? Why is there such a high involvement with misinformation?

Because activists and skeptics are captives of excessive fear.

The excessive fear of the activists is based on their severe concern about the impact of climate change. They believe we are very rapidly heading to our destruction.

Skeptics have severe concerns about the changes being forced into our life. Their concern is elevated because there is a massive involvement of governments in such changes. The excessive fear of skeptics is based on the concern that the shift to low carbon energy–driven by global governments–will be very inefficient. They worry it will lead to energy shortages and very high costs.

Both groups have certain valid concerns. It is important to be fearful to an extent. Some amount of fear is an asset. It can act as a motivator. But excessive fear leads to behavior that is driven by emotions and not by science and realities. Excessive fear allows very little room for critical thinking and is a precursor for poor decisions. It also increases the receptivity to misinformation and increases its spread.

Some activists as well as skeptics suffer from excessive fear because their personal fortunes are tied to specific climate messaging, or fossil fuels or the green energy industry. Examples include financial and political ties and professional success or fame.

The excessive fear of activists and skeptics does not allow them to accept the science and realities of climate and energy. It has created extreme views, which has led to the misinformation war. This war is resulting in a misinformation crisis.

Activists and skeptics are a small fraction of the global population. Most people have not committed to either side. But this is likely to change because the misinformation crisis is exposing the public to excessive fear. The activists and skeptics are using fear as a tool to bring more of the public to their side. This is an effective tool because fear is very contagious. The French philosopher, Michel de Montaigne, said it eloquently[19]. "*There is no passion so contagious as that of fear.*"

The excessive fear caused by the misinformation crisis is likely to affect many more people. We will explore how to resolve the misinformation crisis and target a brighter future in this book.

Book Structure

The common examples of misinformation are discussed in the first part of the book. The topics include climate change, fossil fuels, low carbon power, electric vehicles, and the low carbon energy transition. The misinformation by activists and skeptics for each topic is presented in consecutive chapters to display their contrasting thoughts.

The second part of the book focuses on the big picture. It covers the following topics: crucial facts about climate and energy, robust versus poor policies, how to avoid misinformation and a path to a brighter future.

Certain important points are repeatedly emphasized in the book because they are widely misunderstood.

...§§§-§-§§§...

Misinformation about Climate Change

3. Climate Misinformation by Skeptics: Part I

The climate misinformation promoted by the skeptics is raising doubts about climate science.

What is climate science? It is the collective knowledge from the climate scientists that is well accepted by the general scientific community.

The misinformation by skeptics is dangerous because it is delaying the efforts to address climate change. The amount of climate misinformation that is circulating is gigantic. Here, we will review the misinformation that is not consistent with climate science. Since the list is long, the focus will be on popular examples.

The Earth is Not Warming

Skeptics believe that the earth is not warming. But this belief is wrong.

Some confuse the weather with climate. For example, skeptics use a very cold day as evidence for the absence of warming. Weather is a short-term condition[20]. It fluctuates because of the change in factors such as air pressure, temperature, humidity, and wind speed. In contrast, climate is a long-term average of weather conditions.

So, when skeptics focus on an unusually cold day or week, they are referring to a change in weather. They are confusing climate change with weather change.

Scientists have robust evidence for the long-term increase in global temperatures[21]. **Figure 3.1** shows the change in average global temperature starting from the year 1880[22]. This figure is based on the data collected by weather stations, ships, and buoys from all over the globe[23].

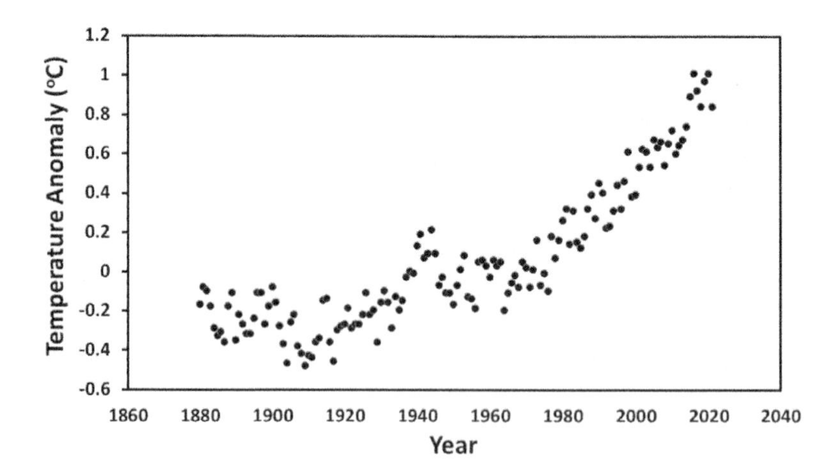

Figure 3.1 The change in global surface temperature compared to the baseline. The baseline is the average of the temperatures from the years 1951 to 1980. Data Source[24]: NASA/GISS

The figure shows a distinct upward trend in temperature over the decades. The data is presented in terms of temperature anomalies. A temperature anomaly is a difference from a baseline temperature. The average of global temperatures from the years 1951 to 1980 is used as the baseline in the figure. The temperature anomaly of 1.02°C for the year 2020 means that year was 1.02°C warmer than the 1951 to 1980 baseline. A temperature anomaly of -0.19°C for the year 1921 means that year was 0.19°C cooler than the baseline.

Skeptics do not trust the temperature data. They believe that the temperature rise is not real because of the large uncertainties. For example, large uncertainties can arise from the different instruments and methods that have been used over time.

Indeed, there is a potential for large uncertainties. However, scientists have robustly dealt with the uncertainties[25,26,27]. They have identified the various sources and minimized the associated uncertainties. For example, scientists have reduced the impact of artificial changes in temperature that arise from the instruments and weather stations. They have done so by using established methods. Also, they have accounted for the uncertainty associated with the removal of the artificial changes.

The results are impressive. For the NASA analysis, the uncertainties (at 95% confidence intervals) are 0.05°C in the annual global average temperatures after the year 1970[28]. Even before 1970, the uncertainty values are not large. The 95% uncertainty is 0.15°C in 1880. There is a larger uncertainty in the earlier years because of an inferior access to quality data.

Scientists have included several checks and balances. Below are examples of the checks and balances used in the NASA analysis[29].

- Adjustments are applied, but the raw data is not changed. Original records are preserved for transparency.
- Climate models are not used in any stage of the analysis.
- All the data is available in the public domain and all the computer code used is available for independent verification.

A good check for reliability is to compare the analysis from research groups which use different methods. Such a check shows excellent consistency. The analyses from the different groups show a similar upward trend in global average temperatures **(Figure 3.2)**[30].

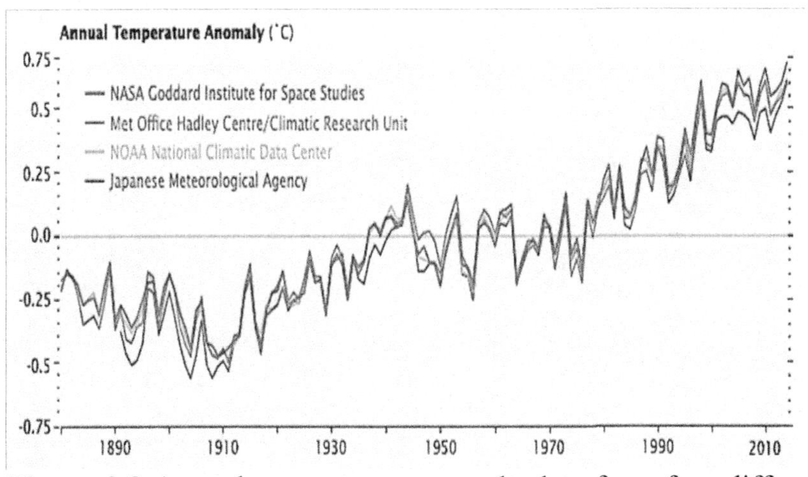

Figure 3.2 Annual temperature anomaly data from four different sources. Data Source[31]: NASA/GISS

Carbon dioxide is Not the Cause

Skeptics believe the CO_2 emitted by humans is not warming our planet. There are several misconceptions behind this belief.

- Some believe that the observed warming is the result of an increase in solar energy received by earth.
- Some point to the very low content of CO_2 and the high content of water vapor in the atmosphere.
- Some believe that the temperature rise can be explained by the urban heat island effect.
- Some also suggest that CO_2 cannot be a problem because plants require CO_2.

We will consider each of these claims.

Is global warming a result of the change in the solar energy received by earth?

If this was true, the change in global temperature would correlate with the change in solar energy over time. **Figure 3.3** compares the change in solar activity over time with the change in global temperature[32].

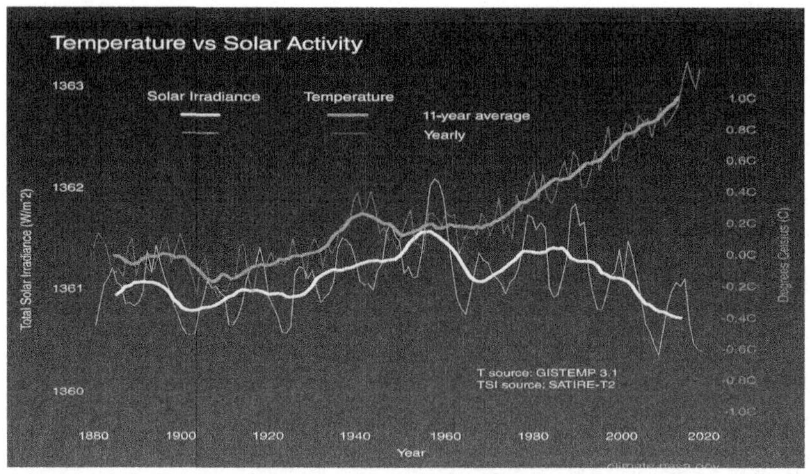

Figure 3.3 Comparison between the changes in the global surface temperature and sun's energy (solar irradiance) received by earth. Data Source[33]: NASA-JPL/Caltech

The solar energy information is presented in terms of solar activity or total solar irradiation. The figure shows that the global temperature has increased distinctly since the year 1960. But there has been no concurrent increase in the solar energy received by earth. In fact, there has been a slight decrease in the solar energy received by earth. This shows that an increase in solar energy is not responsible for the warming.

What is the role of CO_2 and water vapor?

To answer this question, we need to consider the science of greenhouse gases. Water vapor, CO_2, methane, nitrous oxide, and fluorinated gases are the greenhouse gases in our atmosphere. They provide a blanketing effect over earth because of their powerful ability to trap heat. If our atmosphere did not have greenhouse gases, more heat energy would have escaped back into space. This would have resulted in a hostile temperature of -18°C on earth[34]. The temperature is 33°C higher because of the greenhouse gases. Thus, greenhouse gases are responsible for the comfortable temperature on earth.

Earth's heat balance depends on the greenhouse gas content of the atmosphere. Human activities can disturb our planet's delicate heat balance by increasing the greenhouse gas content[35].

Unlike water vapor, the other greenhouse gases are not condensable. These gases are long-lived. For example, some of the CO_2 can remain in the atmosphere for over a hundred years[36]. Human activities produce several times more CO_2 than the other long-lived greenhouse gases. Hence, more emphasis is placed on CO_2.

Water vapor is the major greenhouse gas component in the atmosphere. But it only plays a secondary role in the earth's temperature rise[37]. Let us look at the reasons below.

Water vapor is a condensable gas that can convert back to water. In contrast, CO_2, methane, nitrous oxide, and fluorinated gases are non-condensable, long-lived gases. Human activities release large amounts of these gases. As these gases are long-lived, they accumulate in the atmosphere. This increases their atmospheric content and their blanketing effect. This, in turn, increases the earth's temperature.

What about water vapor content? The temperature of the air determines how much water vapor it can hold. Specifically, air at higher temperature can hold a higher water vapor content.

The rise in earth's temperature because of the long-lived greenhouse gases results in extra water vapor in the atmosphere. The greenhouse effect from this extra water vapor further increases earth's temperature[38]. In other words, the extra impact of water vapor is a result of the accumulation of long-lived greenhouse gases in the atmosphere. The extra impact would not exist if the long-lived gases did not accumulate in the atmosphere.

Let me summarize. The accumulation of long-lived gases increases the temperature content. This, in turn, increases the water vapor content. So, the impact of the additional water vapor only exists because of the long-lived gases[39]. That is why the focus is on the long-lived greenhouse gases.

Complex systems are impacted by many factors. Mathematical models are useful for studying such systems because they can quantify the impact from each of the factors[40].

Our climate system is very complex. Hence, scientists have paid special attention to developing climate models. These models have been used to explain the temperature anomalies discussed earlier[41]. Several factors can influence the temperature. However, the models can only explain the temperature data by allowing for a strong impact of greenhouse gases on temperature. Other factors cannot explain the temperature anomalies[42]. Thus, climate models provide further evidence for the influence of greenhouse gases on earth's temperature.

The CO_2 content in the atmosphere is 0.04%[43]. How does such a small amount have such a large impact?

Because the impact of a substance does not depend only on its quantity. It also depends on its properties. For example, the legal blood alcohol content for vehicle drivers in most countries is between 0.01 to 0.08%[44]. This is because of the properties of alcohol. Even levels of blood alcohol as low as 0.05% can impair human abilities[45]. Similarly, the large impact from greenhouse gases–at very low levels–is because of their powerful ability to trap heat.

What about the urban heat island effect? Is it the cause for the observed warming trend as opposed to CO_2?

An urban area is warmer than the adjacent rural areas. This urban heat island effect is because of some specific features[46]. An urban area has fewer natural landscapes such as vegetation and water bodies that can cool the surroundings. Urban materials such as pavements, roads, and rooftops add to the heat effect because they absorb and emit more of the sun's heat. The large concentration of human activities in an urban area also contributes to the heat effect.

But the urban heat island effect is not the cause for the global warming trend[47]. How do we know that? Because scientists can filter out these effects from the long-term trends. Scientists have known about the urban island heat effect for a long time. They account for the effect in their analysis.

What about plants and the levels of CO_2?

Plants require CO_2 for photosynthesis. Humans have emitted over 2000 billion tons of CO_2 over the past few centuries[48,49]. Studies show that the increase in CO_2 is having a fertilization effect on plants[50]. But plants and oceans can only absorb half of the extra CO_2 emissions[51]. The other half accumulates in the atmosphere. This accumulation is upsetting the heat balance of our planet. It is causing a long-term increase in temperature and shift in weather patterns. The current CO_2 levels are higher than any time in the past eight hundred thousand years[52].

There is No Consensus about Climate Change

Skeptics claim that there is no consensus amongst global scientists. They base their claim on the small fraction of scientists who are skeptical about climate change. As discussed below, this claim is not true.

What is consensus? In practical terms, it is a broad agreement or an agreement between a large majority[53]. So, a scientific consensus means that a large majority of the scientists agree.

Let us first consider the view amongst climate scientists. The goal of climate scientists is to understand critical issues about climate science. Climate scientists publish their findings in scientific journals. So, we can understand their scientific views through their research papers.

Studies have looked at tens of thousands of research papers to understand the view amongst climate scientists. They found that only a small fraction of the papers was skeptical about climate change[54,55]. This analysis has a high level of transparency. Any person can redo this analysis using a free resource such as Google Scholar[56].

Such analyses have shown that there is a broad agreement about climate change amongst climate scientists.

What about consensus in the rest of the scientific community?

Several million scientists are located all over the world. How do we know if there is a general agreement?

Most scientists belong to one or more scientific organizations. The scientific organizations around the globe openly and strongly support that climate change is a serious problem. This includes virtually every national academy and major scientific organization[57,58,59,60,61,62,63,64,65].

Such open and strong support from the organizations is possible because there is general agreement amongst its members about climate change. The support or consensus amongst the general scientific community is not surprising. It is a logical response. It acknowledges the efforts of climate scientists over the past several decades. Climate scientists have documented their efforts in tens of thousands of scientific papers. These papers have provided firm evidence of climate change. It would not be logical or ethical for scientists from other fields to ignore these findings.

Climate science is a very complex field. So, we can expect a few scientists to have different views about climate change.

This leaves us with one key question. Has the minority presented a smoking gun argument?

It has not. So, there is no reason to believe the minority and reject the large majority. It is common sense to accept the joint conclusion from decades of meticulous research from thousands of climate scientists.

In summary, the existence of a small fraction of dissenting scientists does not support the claim that there is no consensus about climate change.

An important note. The consensus amongst the global scientists is that climate change is a serious problem. The consensus does not extend to the highly pessimistic or end-of-the-world views presented by climate activists.

Weather Cannot be Predicted, So how can Climate?

It is a fact that weather cannot be accurately predicted even two weeks in advance. Skeptics use this to argue that it should be impossible to predict the climate many decades in advance. But this argument is flawed.

Weather and climate have major differences. Weather is a short-term condition. In contrast, climate is the long-term average of weather. So, two different types of models are used to predict weather and climate[66].

The differences between weather and climate models are discussed below.

The weather models need to predict the exact temperature, rainfall, etc., for a period at a location. An example would be to predict the exact daytime high temperature after two weeks in Paris.

Climate models are not trying to predict, for example, the exact daytime high temperature in Paris for May 10, 2045. They are trying to predict the average daytime high temperature for the month of May over the entire decade of the 2040s.

The exact weather at a location can vary dramatically on an hour-to-hour or day-to-day basis. In contrast, the average weather varies much less from year to year or decade to decade.

The timescales for the predictions are drastically different for weather and climate models[67]. The goals of the models are very different. Also, the inputs required to predict weather and climate are different. These differences show why we cannot compare weather and climate models.

The challenges and goals of the weather and climate models are markedly different. It is not correct to extrapolate the performance of weather models to climate models. Scientists have access to large amounts of excellent quality historical data. This data serves as input for the climate models. Such models can provide useful information about the expected trends for future scenarios.

One important note about climate models. Climate systems are very complex and there are many unknowns. The term "unknowns" can be described as the gaps in our understanding. The importance of these unknowns in the model predictions rise steeply as we move away from the normal conditions. So, the predictions related to major changes or tipping points have very large uncertainties.

Climate Change is a Hoax

This false information is based on a conspiracy theory. Most conspiracy theories are not consistent with facts and logic. But excessive fear can make people, who are otherwise rational, believe in them.

Some skeptics believe that climate change is not real. They believe it is a hoax created by climate scientists. But this claim is not consistent with scientific facts and logic.

From a scientific viewpoint, there are two key questions. Is climate change consistent with scientific principles? Is it consistent with real world observations?

Recall, climate change results from a heat imbalance caused by the increase in content of the long-lived greenhouse gases in the atmosphere.

Scientists discovered that the greenhouse gases in the atmosphere could capture heat centuries ago. Humans have emitted enormous quantities of these gases since the industrial revolution. The cumulative CO_2 emissions have increased from 0.01 billion tons in 1751 to over 2000 billion tons currently[68].

Scientists have accurate tools to monitor the weather and atmosphere. The global monitoring stations have recorded a large increase in the atmospheric content of greenhouse gases. For example, the atmospheric CO_2 content has increased by over 25%, methane by over 30% and nitrous oxide by over 10% in just the past fifty years[69].

Scientists have also developed specialized tools to understand the climate of the past. Examples of such tools include ice cores, and ocean sediments[70,71]. Studies using the tools have been very revealing. They have shown that the current CO_2 levels in the atmosphere are markedly higher than in the past several hundred thousand years[72]. CO_2 levels were between 180 and 300 ppm for the last eight hundred thousand years before humans started burning fossil fuels.

The heat imbalance from the excess greenhouse gases has caused a rise in the average global temperature. The average global surface temperature has already increased by more than 1°C since the industrial revolution. This is important because the climate is sensitive to even a small change in average temperature. A simple analogy would be to consider the impact of a body temperature change in the case of humans. A decrease (hypothermia) or increase (fever) of even a few degrees beyond the normal body temperature has a significant negative impact[73,74].

As discussed earlier, the rise in average temperature of the earth's surface has been well documented. The rise in temperature has impacted the climate. Climate scientists have gathered evidence about several climate impacts using a vast network of weather monitoring tools **(Figure 3.4)**[75].

Monitoring Tools: World Meteorological Organization

Surface Weather Stations: 10,000+

Upper Air Stations: 1000

Ships: 7000

Weather Radars: 100+

Buoys: 1000+

Special Aircraft: 3000

Meteorological Satellites: 30

Research Satellites: 200

Figure 3.4 Tools available for monitoring the weather and climate. Data Source[76]: WMO

Examples of impacts from climate change include the warming of oceans, shrinking of ice sheets, melting of glaciers, rise in sea levels, acidification of oceans, fertilization effect on plants and an increase in extreme events in many regions[77,78].

Established science and the massive data from the monitoring tools show that climate change is a real and a serious problem.

Moreover, the collusion theory is not consistent with logic.

The climate science community is diverse[79]. It includes scientists from institutions that are spread across the globe[80]. Climate scientists have conducted these studies over many decades. No country, institution, or special interest group has control over the conclusions from the collective studies.

Each country has its own agenda. Despite this, climate scientists from all over the world agree that climate change is a serious problem.

The general scientific community has also accepted the conclusions. Recall, the major scientific organizations across the globe support the common conclusions about climate change. Such agreement amongst an extremely diverse population is only possible if it is based on strong scientific evidence.

Skeptics use the past claims by certain scientists to challenge the conclusions about climate change. Certain scientists have made wild predictions that were completely wrong. For example, the famous physicist John Holdren proposed in the 1980s that the famines caused by climate change could kill a billion people by 2020[81].

It is not appropriate to use such predictions to challenge climate science. The wild predictions were based on speculations. They were subjective and not supported by scientific facts. The global scientific community did not support such predictions.

The situation is different for climate change. The main conclusions about climate change are based on scientific principles and a gigantic number of observations. That is why global scientists accept these conclusions.

I will discuss an example to illustrate the point. Extensive research has shown that obesity can lead to major diseases[82]. Medical science and observations support this conclusion. Therefore, the medical community all over the world accepts the conclusion. Wild claims in the past, present, or future cannot challenge the validity of the strong connection between obesity and long-term health.

Chapter Highlights

Skeptics routinely make factually incorrect claims about the climate. Popular examples of misinformation are listed below:

- There has been no significant warming of earth.
- There has been some warming, but CO_2 is not the cause.
- There is no consensus about climate change amongst the global scientists.
- Since weather cannot be predicted even two weeks in advance, it is not possible to predict climate change in the future decades.
- Climate change is a hoax created for the benefit of certain entities.

...§§§-§-§§§...

4. Climate Misinformation by Skeptics: Part II

Misleading information about the climate has been on the rise over the past decade. Here, we will consider the misleading information that is being circulated by skeptics. The goal of this type of misinformation is to convince the public to delay climate mitigation actions.

Skeptics mislead mainly by omitting crucial information.

Climate Has Always Been Changing

A common argument by the skeptics is that we do not need to worry about climate change because it has always been changing. But as discussed below, this argument is flawed.

Our climate has been changing since the beginning of time[83]. But there are two major differences between the past and the current change in climates.

First, humans are driving the current change in climate. So, we can mitigate its impact. In contrast, the past changes were driven by natural causes[84].

Second, the nature of the recent change is unusual compared to the past. As discussed below, the speed of the recent warming is markedly faster.

Earth's temperature has swung between cool periods and relatively warm periods many times over the past two million years[85]. During the shift from the last cool period to the current warm period, the temperature increased by 5°C. This change was very slow. It took place over 5000 years. For comparison, earth's temperature has warmed by more than 1.1°C over just the last 150 years. Robust data is available which shows that the speed of the warming in the last 50 years has been more than any other 50-year period in the last 2000 years.

Below are two more examples that illustrate the unusual speed of the current impact.

- The recent acidification of the oceans is far faster than during the past several million years[86]. Sea levels have also been rising faster. The rise in sea levels since the year 1900 has been faster than any other century in the past 3000 years[87].
- The frequency and intensity of extreme events such as heat waves have been increasing at a fast pace in the recent decades[88]. For example, the frequency of heat waves per year in the United States have increased by a factor of three since the 1960s[89]. Heat waves are of concern because of their impact on health, agriculture, and energy production.

Slower rates of change, like in the past, allowed for a long time for the species to adapt to the changing climate. The much higher speed of the recent change is a serious concern.

Climate experts agree that the impact from climate change is going to worsen significantly with time[90]. This is consistent with the trends observed over the past decades and the output from climate models. The more we wait, the worse will be the impact. So, urgent action is necessary.

The term urgent action can have a different meaning for different people. I will take a moment to share my definition. Urgent action is not a rapid embrace of inefficient solutions. Instead, urgent action involves a well thought out approach that will enable the fastest possible sustainable transition.

We Can Manage the Impacts of Climate Change

Skeptics use two main arguments to support this misinformation.

The first argument uses the decrease in deaths from climate disasters as evidence to suggest that we can manage the impacts. Examples of climate related disasters are shown in **Figure 4.1**.

Examples of Disasters

- ➢Extreme Temperatures
- ➢Droughts
- ➢Floods
- ➢Glacial Lake Outbursts
- ➢Storms
- ➢Landslides
- ➢Wildfires

Figure 4.1 Examples of natural disasters that can be impacted by climate change.

The global deaths related to climate disasters have decreased markedly over the last decades[91]. The deaths have fallen by a factor of three from the decade of 1970s to the 2010s. During this time, the average global temperature has increased by 0.9°C[92].

The number of deaths has decreased despite the fourfold increase in the number of climate disasters over that period[93]. The deaths have mainly fallen because of an increase in the use of early warning systems. The early warnings have enabled evacuations in a timely manner and have prevented many deaths.

But deaths are not the only measure of the impact of climate disasters. The number of people affected, and the economic losses have increased over the decades. Examples of how people are affected include injuries and loss of property and livelihood.

The climate disasters have caused large disruptions in many lives and significant economic losses. For example, climate disasters affected 1.7 billion people and caused damage of $1.4 trillion during the last decade[94,95,96]. The impact was much lower in the 1970s.

The number of climate disasters have increased over the decades because of climate change. Climate experts agree that the frequency and intensity of climate disasters will continue to increase with an increase in warming[97]. This will further increase the number of people impacted, and the economic damages. The impact of climate disasters is already substantial and is expected to worsen with time.

The higher the temperature rises, the higher will be the impact on lives, and our ecosystem. It is misleading to argue that the decrease in deaths from climate disasters is evidence for our ability to manage the long-term impacts from climate change.

The second argument uses climate adaptation as an excuse. Climate adaptation involves the anticipation of the adverse effects of climate change and taking action to minimize damage[98,99]. Examples of adaptation are shown in **Figure 4.2**.

Examples of Climate Adaptation

- **Developing early warning systems**
- **Building climate resilient infrastructure**
- **Restoring ecosystems**
- **Ensuring water security**
- **Ensuring energy security**

Figure 4.2 Examples of climate adaptation.

Climate adaptation is a crucial part of the solution. It can drastically decrease the life, property, and livelihood loss that is caused by climate change. But it is not a standalone solution for addressing climate change.

There is scientific consensus that the impacts of climate change will increase rapidly if we do not take efforts to mitigate it. If we only focus on adaptation, we will always be playing catch up. The more we delay our efforts to mitigate climate change, the more we will have to adapt. This will greatly increase the technical and financial challenges for a good outcome.

The impacts of climate change are varied and affect every aspect of our ecosystem. Also, the risk of very high impact, low probability events increase with increasing warming.

Climate adaptation by itself will fall short because of technical and financial limitations. This will cause a severe impact on the global society and every aspect of our ecosystem. A robust path forward must include urgent efforts to mitigate climate change along with adaptation.

Climate Science is Not Settled

Skeptics suggest that climate change cannot be viewed as a serious problem because there are many uncertainties associated with climate science. This is a misleading argument.

Our climate is an extremely complex system. It consists of five components: a) atmosphere, b) biosphere, c) cryosphere, d) hydrosphere, and e) land surface[100].

These components interact with each other in a convoluted manner. Uncertainties in climate science are expected because of the unusual complexity. The uncertainties will never be eliminated. But efforts that improve the understanding about the climate can decrease the uncertainties.

Over the decades, climate scientists have put in massive efforts to decrease the uncertainties[101]. They have collected a large amount of quality data using a vast network of weather and climate monitoring tools. They have used sound scientific principles to gain insights from the data.

Because of these efforts, there is high certainty about several aspects of climate change and its impacts. We will review below some key aspects about our climate that are understood with high confidence and have global relevance[102]. The information is from the latest report by the United Nations body on climate change (IPCC)[103].

- The increases in CO_2 and methane concentrations since 1750 are much higher than any natural changes over at least the past eight hundred thousand years.
- Human influence has warmed the atmosphere, ocean, and land. The global temperature has increased faster since 1970 than in any fifty-year period over at least the last two thousand years.
- Extreme events such as heat waves have become more frequent and intense across most land regions since the 1950s. Humans are responsible for these events.
- Humans are contributing to the global retreat of glaciers, decrease in artic sea ice and Greenland ice sheets.

- The global mean sea level has risen faster since 1900 than over any prior century in at least the last three thousand years. Humans are responsible for the rise.
- Humans are responsible for the acidification of oceans.
- The changes are negatively impacting our terrestrial and ocean ecosystems. This is critical because we are strongly connected to our ecosystems.
- Over three billion people live in regions that are highly exposed to climate change. A high proportion of other species are also exposed to climate change.
- Greenhouse gas emissions cause several changes that are not reversed over hundreds or even thousands of years. Examples of such changes include those in the oceans, ice sheets and sea levels.
- The impact associated with climate change is expected to increase significantly with every incremental increase in global warming.

The scientific community has high confidence in the aspects discussed above.

A large amount of quality data is required to gain high confidence. Such data is not yet available for all the climate impacts. Because of the lack of adequate data, several of the climate impacts are currently understood at low or moderate confidence levels[104]. As more data becomes available, more of these impacts will be understood with high confidence.

But one fact is key to this discussion. The impacts that are known with high confidence are already adequate to show that climate change is a serious problem.

Many aspects of climate science are not yet completely understood. This is because of the ultra-complex nature of the topic. But several aspects of the climate are well understood, i.e., scientists have a high confidence in these. This settled knowledge is adequate to show that climate change is a serious problem. It is misleading to use the excuse that climate science is not settled to delay mitigation efforts.

An important note. Using inefficient solutions can also cause serious problems. So, it is crucial that the mitigations efforts are well thought out.

Chapter Highlights

Skeptics routinely make misleading claims about the climate. Popular examples of misinformation are listed below:

- We do not need to take urgent action because the climate has always been changing.
- We can manage the impacts of climate change.
- We do not need to take urgent action because there are too many uncertainties about climate change.

...§§§-§-§§§...

5. Climate Misinformation by Activists

Previously, we looked at examples of climate misinformation that are being spread by skeptics. Here, the focus is on the misinformation by activists. This misinformation is causing an excessive fear about climate change.

The alarmism from activists is dangerous. It is causing needless mental anguish. Also, it is making it far more difficult to address climate change in a sustainable manner.

Here, we will explore the nature of the climate misinformation by activists. The misinformation mainly involves the use of poor assumptions, and omission of crucial information.

Impact of Climate Disasters is Increasing Very Rapidly

This is a very popular claim made by activists. It is causing extreme fear amongst many in the global population. As discussed below, this claim is not supported by facts.

I will first share some background information. Climate change is increasing the frequency and intensity of certain natural disasters[105]. Higher the frequency and intensity of the disasters, higher is the impact on the global society. The disasters that climate change can affect include floods, extreme temperatures, droughts, wildfires, landslides, floods, and glacial lake outbursts.

Activists believe that climate change has caused the impact of these disasters to rise very rapidly in recent years and decades. Many believe that each year is far worse than the previous in terms of climate impact. And every decade is far worse than the previous decade.

They believe that the disasters have been causing unprecedented havoc on global society in recent years. They are convinced that these acute impacts have made our planet a scary place.

Climate change is a serious problem and is having a significant impact. But are we in a scary place? Is it true that the global impact from the disasters has skyrocketed in recent years and decades?

I will share global records to discuss how the impact of disasters has changed with time. Specifically, I will share global data about three critical indicators. These indicators tell us about the impact of the disasters.

- The first indicator is the number of deaths from disasters.
- The second indicator is the number of persons affected by the disasters. This includes the number of persons injured, those rendered homeless, and those requiring help of any kind.
- The third indicator is the economic damage caused by the disasters.

The basic data about these indicators is made available by the emergency events database (EM-DAT: The International Disasters Database)[106,107]. It is the most widely used database on global disasters. The United Nations, World Meteorological Organization, and academia are examples of some of the users[108].

First, let us consider the data on the global deaths. There has been a drastic decrease in deaths from disasters. Global deaths have decreased by a factor of three over the past five decades. The improvements in early warning systems have been the main reason for the decrease[109].

Next, we will consider the global trend for the number of persons affected by disasters[110]. The number of affected persons has increased over the decades **(Figure 5.1)**[111].

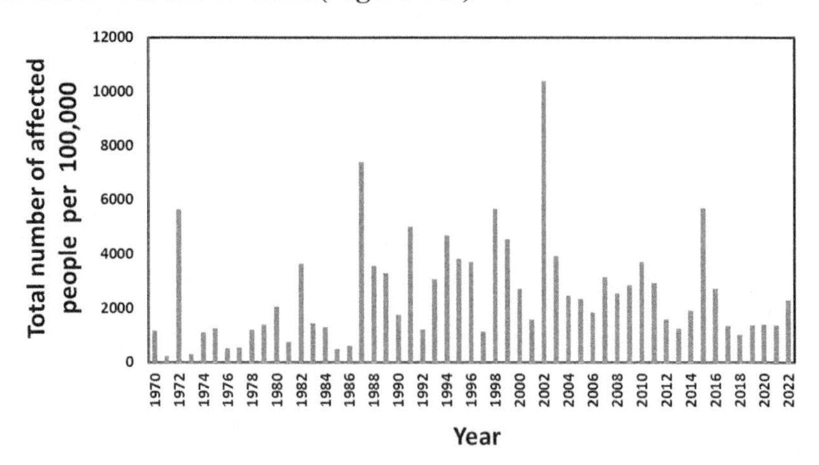

Figure 5.1 The total number of persons affected per year by climate related disasters per 100,000 of the total population. This data is provided per 100,000 people to account for the change in the

global population size over time. The total number affected is the sum of injured, homeless, and those requiring any type of assistance. Data Source[112]: Our World in Data (EM-DAT, UN world population prospects 2022).

This data shows that climate change has had a significant impact. But it also shows that there is no reason to panic. The number of people affected globally has not increased more rapidly in recent times[113].

Finally, we will consider the economic damage caused by the disasters. Gross Domestic Product (GDP) is widely used as a gauge of the economy. Therefore, it is logically robust to discuss the economic damage in terms of %GDP. The global impact from the climate disasters has increased over the decades **(Figure 5.2)**[114]. But the increase has not been unusual in the most recent years. The data on economic losses does not show a scary trend.

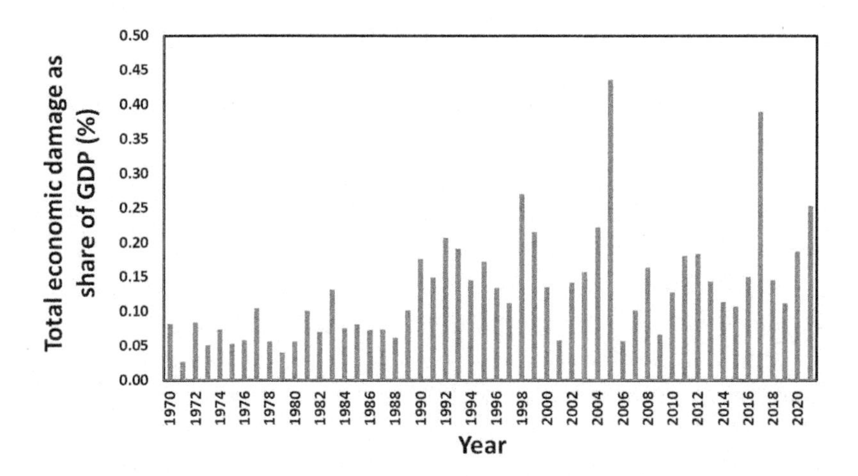

Figure 5.2 Economic damage caused by climate related disasters by year as a share of GDP. Data Source[115]: Our World in Data (EM-DAT, World Bank)

Climate change is a serious concern, and we must address it. But there is no cause for panic based on the impact data from the disasters. For example, less than 3% of the global population has been affected by the disasters annually in recent years. Also, the average impact on the global economy has been less than 0.2% of the GDP. On a global level, the recent impacts from climate related disasters on the number of people affected and economy are not too different from the 1990s and 2000s **(Figures 5.1 and 5.2)**.

What is the reason behind the panic or sense of alarm?

Media has greatly increased the coverage of disasters in recent years. Disasters cause a lot of devastation. Images and videos of the property damage and human stories about their losses are very disturbing. We are now being constantly bombarded with such videos and images. Each disaster is hyped to be one of the worst ever. This massive media coverage makes it seem as if the recent years have been very scary and unusual.

Also, disasters are often discussed in a misleading manner. Misleading by omissions is very common. Activists use limited data and present it in a manner that makes the current state look scary.

I will discuss two examples.

The first example borrows from my earlier discussion. Activists do not discuss the trend in global data as shown in Figures 5.1 and 5.2. They focus on regional data and select timeframes (e.g., certain regions and years) to hype the devastation. They only showcase selected data that supports their narrative of doom. This has been creating a false perception about the current state.

The second example is about how activists are using selected data from research papers to support their narrative. Suboptimal temperatures can cause excess deaths. These excess deaths are those that occur above and beyond normal conditions. A recent research paper has discussed the excess deaths because of suboptimal hot and cold temperatures[116,117]. The paper has four key conclusions based on data over the past two decades. 1) The excess deaths because of suboptimal hot temperatures have increased over the past two decades. 2) The excess deaths because of suboptimal cold temperatures have decreased. 3) The overall excess deaths because of suboptimal hot and cold temperatures have decreased. 4) Suboptimal cold temperatures have contributed to nine times more excess deaths than suboptimal hot temperatures.

Some activists are misusing the research paper to promote alarmism. They use the tagline: new research shows that hot extremes are causing record high deaths. They omit the three key conclusions that oppose their narrative.

We Could Soon Fall Off a Cliff

Activists overstate the impact of climate change. Poor assumptions lead to alarmism.

Activists believe that the global temperature will soon reach a level that will cause major destruction across the globe. They believe that survival will shortly be an issue for the global society.

The current rise in temperature or warming above the pre-industrial levels is 1.1°C [118]. Some believe that a warming of 1.5°C is the limit, while others believe it is 2°C and so on. Activists believe that human society will fall off a cliff once that limit is exceeded. They believe that our society will collapse at that point.

As discussed earlier, activists believe we are already in a scary state. This makes it easy for them to believe that we could soon fall off a cliff.

Scientists agree that the climate impacts will worsen markedly with an increasing rise in global temperature[119,120]. But they do not see us falling off a cliff in the foreseeable future.

To illustrate this point, we will review the predictions about the impact of warming on the frequency of certain extreme events. These predictions are from the latest IPCC report[121].

I will discuss the data relative to our present state. Presently, the warming is 1.1°C. The frequency of events per ten years are discussed. Compared to the present, the frequency of extreme temperature events is expected to increase by a factor of 1.5, 2.0 and 3.4, respectively, for warming scenarios of 1.5°C, 2°C, and 4°C, respectively[122]. Heavy precipitation events over land are expected to increase by a factor of 1.2, 1.3, and 2.1, respectively, for the three scenarios[123]. Droughts are expected to increase by a factor of 1.2, 1.4, and 2.4, respectively, for the three scenarios[124]. The impact on the ecological system is also expected to be markedly higher with a higher rise in temperature.

The predictions indicate a much worse impact for a 4°C warming scenario. But there is no consensus amongst the scientists about a cliff at 4°C. There is no reasonable evidence for the existence of such a cliff.

How much warming can we expect by the end of this century?

The United Nations recently released a report that addresses this question[125,126]. The report estimates that the temperature will rise between 2.5°C to 2.9°C based on the current policies and Paris pledges. Even with the current policies and pledges, the rise will be far lower than 4°C by the end of this century. There is no reason to believe that a warming between 2.5°C to 2.9°C will cause us to fall off a cliff.

The International Energy Agency (IEA) estimates a temperature rise of 2.4°C in their stated policies scenario, which considers existing policies and those under development[127]. Global governments are expected to improve over their current policies. It is reasonable to expect that the warming will be lower than 2.5°C by the year 2100. Clearly, there is no reason to panic.

What about the impact on the economy?

Any abnormal influence has a direct impact on the economy. The COVID-19 epidemic is a recent example. It caused a loss of six percentage points to the global economy in 2020[128,129,130]. So, the impact on the global economy can be used to evaluate the seriousness of the climate impacts.

Researchers have estimated the future economic impact of climate change. The studies provide a wide range of economic estimates depending on the assumption used[131,132,133]. Even under assumptions that are pessimistic, the estimated impact is far from extreme. Examples of pessimistic assumptions are high warming levels (≥ 4°C) and limited climate adaptation. Under realistic assumptions, the future global economic impacts are modest[134,135,136]. Again, there is no reason for panic.

Some regions will be affected far more than others. The future impacts in these regions could be very severe in the absence of climate adaptation. But adaptation efforts in a timely manner can greatly reduce the worst of the impacts in such regions.

So, why are so many activists in a panic mode?

Activists consider the worst-case scenarios when discussing climate impacts. Also, they discount adaptation efforts to paint a cataclysmic picture. I will discuss two examples.

While discussing the increase in sea levels, activists focus on the impacts that would arise from a ≥ 4°C warming by the year 2100. Others assume coastal cities will not take adaptation steps to protect themselves. Such assumptions can lead to the alarming conclusion that many cities could be under water by the end of this century. Realistically, the warming by the end of the century is expected to be less than 2.5°C[137,138,139]. The increase in sea levels at this level of warming will be much lower than at 4°C warming. Also, the endangered cities will very likely take the necessary adaptation steps. Effective examples of adaptation include deployment of dikes, embankments, sea walls or surge barriers. Netherlands was able to deploy an effective solution for this problem decades ago[140]. A third of the Netherlands is below the sea level[141]. Dikes have provided excellent protection from the sea for the Netherlands.

Activists are alarmed about very high-impact events that can be caused by climate change. These are events where the change is massive and abrupt. An example is the runaway warming that could be caused by an abrupt release of methane trapped in the ocean floors. Activists believe such events will occur within the next decades and cause untold destruction. They believe so based on some speculative modeling.

But it is not practical to believe these speculations. No climate model can predict if or when a massive and abrupt event will occur. This is because of the many unknowns and the lack of relevant data. It is impossible to make credible predictions when there are many unknowns, and the data is extremely limited.

What is the consensus amongst global scientists? They agree that the risk from such events would significantly increase with the increasing warming of the planet[142]. But most agree that the probability of these events is low in the foreseeable future.

A proper context is crucial. There is no reason to believe that such climate events pose a higher risk than a global war or the outbreak of an extremely dangerous disease. So, the alarm about extreme events is overblown. We must take urgent actions to lower the risk, but there is no reason to panic.

Climate Mitigation Is Our Most Urgent Issue

Activists often hype that climate change is our most urgent issue. They believe that the mitigation of climate change deserves an unconstrained allocation of resources.

Does climate mitigation deserve the highest priority?

Who decides this? Is it the scientific community, or representatives from the United Nations, or think tanks?

A small fraction cannot make this decision. The answer depends on what most of the global population believes.

To address this question, we will review the key global surveys undertaken on this topic over the past decade.

The United Nations reported the results from a major survey on this topic in December 2014[143]. It was the United Nations largest and most inclusive global survey. More than seven million people across the globe took part. The participants had the required diversity. They represented a very wide range of economic and educational backgrounds.

The survey asked a simple yet very effective question. The question was: "what matters most to you?"[144]. The participants had to choose six issues that were most important to them amongst a list of sixteen issues. Examples of issues were education, political freedom, equality between men and women, and climate change.

The survey identified "a good education", "better healthcare", and "better job opportunities" as the three most important issues. The issue "actions taken on climate change" received the least votes. Fifteen other issues were ranked as more important than the actions taken on climate change.

The United Nations was involved in another large global survey in 2020[145]. It had over a million participants from diverse backgrounds. The survey was undertaken in the early days of COVID-19. Unfortunately, the structure of the questions made it less straightforward to interpret what mattered most to the participants.

The survey asked three key questions. 1) What should the international community prioritize to recover better from the pandemic? 2) Taking a longer view, if you picture the world you want in 25 years what three things you would most want to see? 3) Which of the global trends will most affect our future?

For question #1, the top vote was for "access to basic services: healthcare, water, sanitation, and education". For question #2, the top vote was for "more environmental protection". While for question #3, the top vote was for "climate change and environmental issues".

The survey found that "access to basic services" was the most important issue in the short-term. While environmental and climate issues were most important from a long-term viewpoint.

Another recent survey by the United Nations was specifically about climate change[146,147]. They surveyed over a million people. The global survey showed 55% public support for climate policies from high income countries. But the support was only 36% from the least developed countries. Globally, there was only a 38% public support to urgently mitigate climate change[148].

Ipsos, a multinational market research and consulting firm, recently undertook a global survey about climate change[149]. The survey showed that about two-thirds of the global population believed that their countries should do more to address the climate challenge. But only 30% of the global average said they would pay more taxes to combat climate change.

Another recent global survey by Ipsos was very revealing[150]. The survey asked about the key worries for the population. The top five key worries were inflation, poverty and social equality, crime and violence, unemployment, and corruption. Climate change was ranked eight.

Overall, the surveys show that the current standard of living is the most urgent issue. Despite the progress over the last decades, the living conditions are still very poor for a large fraction of the global population.

I will share the recent information reported by the United Nations[151].

Inadequate access to basic water services was a major issue in 2022[152]. Safely managed drinking water was not available to two billion people. Also, over three billion people had no access to safely managed sanitation. Based on the current progress, the United Nations estimates billions will lack such basic water services even in 2030.

Access to nutrition was also poor in 2022[153]. Over 700 million people suffered from hunger and over two billion did not have adequate access to food. Access to education was also a major issue[154]. Over 40% of the youths did not complete upper secondary school.

Access to adequate and clean energy is critical for a good standard of living. But this continues to be an enormous problem[155]. 675 million people had no electricity in 2021. Over two billion people used highly polluting cooking fuels.

Major issues exist with access to healthcare as well[156]. A significant fraction of the global population people had no access to vital health services in 2022. Five million children died before their fifth birthday because of poor conditions[157].

Let us consider the impact of climate disasters for reference. In the last decade, climate disasters have affected less than 200 million people per year on average. The average number of deaths has been less than 0.05 million per year.

Currently, the lack of access to basic services is causing far more suffering than climate change. Billions are suffering and millions are dying.

A difficult life is not limited to the most distressed fraction of the population. It is widespread. The living conditions are far from adequate for a major fraction of the global population. According to data from the World Bank, 4.5 billion people live on less than $10/day[158]. This meager amount is forcing the population to choose between basic needs such good education, adequate housing, or treatment of health conditions. More than half of the global population is struggling to meet their basic needs. They are in substantial pain. This pain prevents them from being overly concerned about the future impacts from climate change. Climate mitigation is not their most urgent issue. Poor living conditions are a far bigger issue for them.

The affluent population can easily meet their basic needs. They are not suffering today. So, they can afford to worry about future climate impacts.

Yet climate change is not the topmost concern for all in the affluent category. Some are more concerned about the ballooning national debts, energy security, spiking polarization, erosion of confidence in elections, or rising influence of autocracies. Any of these issues can cause extreme disorder in the future.

United States is one of the most affluent nations. I will share recent surveys to discuss the voice of its citizens. Dealing with climate change was a top priority for only 37% of the people surveyed[159]. It ranked a lowly 17 in a list of 21 issues. Another survey about climate-related spending was also very instructive. A key question in the poll was to identify the support for a law that would increase the average monthly cost by a certain amount to combat climate change. Even a small $1 additional cost per month was only supported by 38% of the people surveyed[160,161].

Clearly, the global majority does not believe that climate change is our most urgent issue.

Chapter Highlights

Activists routinely spread misleading information about the impact of climate change. Their information is far too pessimistic. Popular examples of misinformation are listed below:

- The impact of climate related disasters has been increasing very rapidly in recent times.
- We are on the brink of global devastation because of climate change.
- Actions to mitigate climate change deserve the most urgency because climate change is the global society's most urgent problem.

...§§§-§-§§§...

Misinformation about Fossil Fuels

6. Fossil Energy: Misinformation by Activists

Misinformation about climate change is only rivalled by misinformation about energy. Activists have several misconceptions about fossil fuels, which they spread routinely. Some of their misinformation can cause intense fear or anger about fossil fuels. This can make it difficult to select a rational path.

Let us start with a brief background about fossil fuels.

Coal, natural gas, and crude oil are fossil fuels. Scientists have traced their origin to the dead organisms from the distant past[162]. Dirt and water covered the organisms after their death. Heat and pressure–applied over millions of years–converted the remains of the dead organisms to fossil fuels. Coal was formed from dead plants while natural gas and crude oil were formed from dead marine organisms[163].

Fossil fuels have vast reserves because they were formed from plants and marine organisms that were abundant on earth. The three fossil fuels are distinctly different in terms of their physical and chemical properties. Coal is a solid, crude oil exists in the liquid form and natural gas occurs as a gas. The specific properties of fossil fuels define their suitability for different applications.

Fossil fuels have provided most of the global energy for more than a century. Fossil fuels provide over 80% of the global energy currently[164]. Coal and natural gas are mainly used for producing electricity and heat. Crude oil products such as gasoline and diesel are used in the transportation sector.

Misinformation about fossil fuels can lead to poor energy policies and increase energy poverty. It is important to move past this misinformation.

Climate Impact of Fossil Fuels Has Been Understood Since a Long Time

Activists claim that the climate impact of fossil fuels has been well understood for many decades. Some even claim that the timeframe extends over a century. Based on such claims, it may seem that major efforts should have been taken to reduce use of fossil fuels a long time ago. But as discussed below, these claims are not supported by facts.

A scientific topic needs to be well understood to initiate societal actions that involve a radical change.

When is a scientific topic well understood? Only after there is concrete proof, i.e., proof that cannot be refuted.

Proposing a theory is not enough. The theory needs to be validated by obtaining concrete proof. Why? Because new findings have invalidated many theories. More than hundred million research papers have been published over the decades[165]. Novel theories or concepts have been proposed in a significant fraction of these papers. Over time, many of these theories have been proven to be completely or partially wrong. Massive amounts of resources would have been wasted if we had taken actions before the theories were validated. So, concrete proof is a crucial requirement before making a major change.

Obtaining concrete proof for the theories proposed for complex systems is very difficult. It requires major scientific advances.

Was concrete proof about the fossil-based climate impact available for a long time? Historical publications and data show that concrete proof has only been available since the last couple or so decades. The timeline is discussed below.

Svante Arrhenius, a Swedish chemist, was the first to propose that the CO_2 from fossil fuels would cause a temperature rise[166]. He proposed this theory over a hundred years ago. Concrete proof for his theory required an observed rise in global temperature that scientists could distinguish from the natural factors. For example, solar radiation is a natural factor that can change the temperature. Back then, there was inadequate temperature data, no CO_2 data, and little-to-no knowledge about the variation from natural factors. There was no data about the impact of climate change. So, scientists ignored his work.

Callendar, an English engineer, proposed that fossil fuels were causing warming in the 1930s[167]. Again, this was just a theory. The supporting data and knowledge were missing.

Both Arrhenius and Callendar believed that the temperature rise from fossil fuels would be beneficial for mankind[168,169]. This is a key point. These pioneers did not oppose the burning of fossil fuels based on their research. Quite the opposite, they believed that there would be benefits. This shows that until the year 1940, the theory about the impact of warming was that it would be beneficial.

Scientists became more interested in this topic in the 1950s. They made advances in atmospheric CO_2 measurements and several other aspects over the next couple of decades[170,171].

Despite the advances, concrete proof was still missing in the 1970s. In fact, a few scientists even believed that global cooling could be a problem[172,173].

It is helpful to see how the historical temperature data would have looked to the scientists back then. They would have seen no clear trend **(Figure 6.1)**[174].

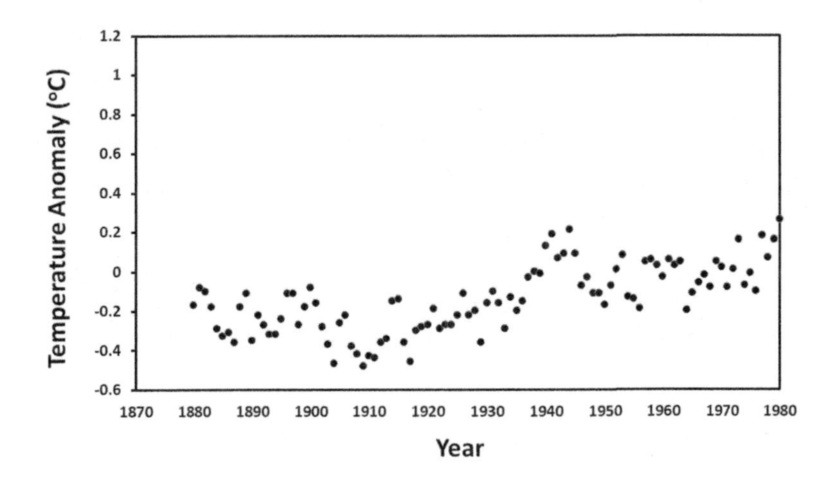

Figure 6.1 The change in global surface temperature compared to the baseline. The baseline is the average of the temperatures from the years 1951 to 1980. Data is shown from the year 1880 to the year 1980. Data Source[175]: NASA/GISS

Specifically, they would have observed the following. No significant change from 1880 to 1900. A decrease in the temperature from 1880 to 1910. An increase from 1910 to 1940. And then a plateauing until 1980.

There were two major challenges for obtaining concrete proof back then. There was no clear trend in temperature data. Also, there were many gaps in knowledge about the climate system. Recall, concrete proof would require a temperature rise that could be assigned only to fossil fuels. This required excellent knowledge about how natural factors affected the climate system.

Majority of climate scientists in the 1970s believed that global warming could become a significant problem[176]. But the required proof to make concrete claims was missing. There were many knowledge gaps.

A prestigious report published in 1975 captured the major gaps in the knowledge[177]. The special report published by U.S. National Academy of Sciences was titled "Understanding Climate Change". It included the following critical comments.

- "Our knowledge of the mechanisms of climate change is at least as fragmentary as our data. Not only are the basic scientific questions largely unanswered, but in many cases, we do not yet know enough to pose the key questions."
- "How do we separate the genuine climatic signal from what may be unpredictable noise, and to what extent are noise and signal coupled? These are important questions and the ones to which there are no ready answers."

Over the next decade, climate scientists undertook major efforts. They addressed many gaps. For example, scientists were able to accurately estimate the contribution to temperature changes from natural factors[178].

Globally, experts became more concerned about climate change because of these studies. The Intergovernmental Panel on Climate Change (IPCC) was established in 1988 to guide global governments[179].

But concrete proof was missing even in the early 1990s[180]. The judgement section from the first IPCC report confirms this. Here are a few statements from the report as evidence for the absence of concrete proof.

- "Global mean surface air temperature has increased by 0.3°C to 0.6°C over the last 100 years, with the five global average warmest years being in the 1980s."
- "The size of the warming is broadly consistent with predictions of climate models, but it is also of the same magnitude as natural climate variability."
- "The unequivocal detection of the enhanced greenhouse effect from observations is not likely for a decade or more."

We can confirm this information from the historical temperature data from NASA[181]. The temperature change seen until the early 1990s was in the same range as variation from natural climate factors. According to the recent IPCC report, natural factors have caused a temperature change between -0.23°C and +0.23°C since 1850[182].

As predicted in the first IPCC report, it took until the end of the century to unequivocally detect the temperature impact from CO_2. Thus, concrete proof for the impact from fossil fuels was available only at the end of the twentieth century[183].

Natural Gas is Not an Option

Activists believe that natural gas is not an option to decrease greenhouse gas emissions. This is a false belief, as discussed below. Replacing coal with natural gas is an important option in the intermediate term for several reasons.

I will discuss five reasons for the confusion about natural gas.

The first reason is that natural gas power plants are often grouped with coal power plants. Most people know that coal plants are large emitters of greenhouse gases and air pollutants. The grouping of natural gas power plants with coal power plants makes it seem that the two are similar, i.e., both are equally bad.

But natural gas power plants are very different. All credible sources agree they are far superior[184,185.]

They emit 50% lower greenhouse gases compared to coal power plants on a lifecycle basis[186,187]. For reference, currently electric cars also reduce greenhouse gas emissions by 50% on an average when they replace conventional cars[188,189].

Natural gas power plants emit far fewer air pollutants compared to coal power plants[190]. Moreover, natural gas power plants have markedly lower upfront costs[191]. Upfront cost is the initial cost to deploy a technology. It is the cost to build a power plant or purchase a car.

Natural gas power has one of the lowest upfront costs in the power sector[192]. It requires a very low initial spending to reduce greenhouse gas emissions[193]. Natural gas power is also competitive in terms of lifetime cost of electricity production in regions with low to moderate natural gas prices[194]. Lifetime cost is the total cost over the whole life. It includes the upfront cost and other costs such as the cost to operate and maintain the technology.

Also, natural gas plants provide high flexibility[195]. This means that they are well equipped to match the electricity supply with demand. Replacing coal power with natural gas power can increase the overall grid flexibility. Replacing coal with most other technologies will reduce grid flexibility. Flexibility is important to maintain a highly reliable electrical grid. High reliability equals fewer disruptions for meeting our 24x7 electricity needs.

Natural gas plants can further reduce CO_2 emissions by using carbon capture and storage (CCS) technology. The CCS technology is not yet cost-effective. But it is expected to improve over the next decade. When it improves adequately, the natural gas power plants can be retrofitted to further reduce greenhouse gases. This can reduce greenhouse gases by an additional 70%[196]. Thus, replacing coal with natural gas is aligned with a low carbon future.

Some methane gas is lost as leakage (aka fugitive emissions) when natural gas is produced and transported. This leads to the second reason for the false belief about natural gas. Activists have an incorrect notion about the impact from methane leakage. They believe that the gas leakage negates the advantage of natural gas power.

This is not true.

Two credible sources–U.S. National Renewable Energy Laboratory and United Nations Economic Commissions for Europe–have provided updates about greenhouse gas emissions from the power sector[197,198]. These studies include the impact from the gas leakage. Both confirm that replacing coal power with natural gas power will decrease greenhouse gas emissions by 50%.

Moreover, the gas leakage can be cost-effectively reduced via policies[199]. Targeting the leakage can boost the advantages of natural gas.

The third reason for the false belief of the activists is the fear about widespread water contamination. The fear is related to the natural gas production process called fracking, aka hydraulic fracturing.

Is the fear valid? How large is the impact on water quality? Let us look at the conclusions from the most reliable sources available.

The U.S. Environmental Protection Agency (EPA) published a report in 2016 about the impact of fracking on water supply[200]. The report concluded that fracking could impact water quality under certain circumstances. They identified the circumstances so that the producers had the information to minimize water contamination.

The following highlight from the U.S. EPA website confirms that they have no overall concern[201]. "Unconventional oil and natural gas play a key role in our nation's clean energy future. The U.S. has vast reserves of such resources that are commercially viable because of advances in horizontal drilling and hydraulic fracturing technologies. These technologies enable greater access to oil and natural gas in shale formations. Responsible development of America's shale gas resources offers important economic, energy security, and environmental benefits."

The EPA website basically conveys that fracking when conducted properly has many key benefits.

The U.S. Geological Survey (USGS) also found that there was no reason for the alarm. USGS is the largest water, earth, biological science, and civilian mapping agency in the United States. Specifically, the agency collects, monitors, analyzes, and provides scientific information about natural resources, issues, conditions, and problems[202].

Natural gas has been produced by fracking in the United States for over two decades[203]. Consequently, the USGS is in a unique position to study issues related to fracking. Over the years, the USGS has dedicated large resources to study the impact from fracking[204,205]. Their studies show fracking has not caused widespread contamination of water supplies[206,207,208].

The USGS studies show that fracking is reasonably safe when operations are conducted properly. Expectedly, incorrect operations can cause problems. This is a challenge for most industrial operations and is not limited to fracking. Also, USGS has found no evidence for major concern from earthquakes related to fracking or associated operations in the United States[209]. So, there is no cause for alarm.

The fourth reason is the myth that if we do not instantly stop the temperature rise, we will shortly drop off a cliff. This leads to the belief that use of all fossil fuels must be stopped instantly. The myth of an impending cliff has been debunked earlier.

The fifth reason is the false belief that natural gas power is not required. The belief is that clean alternatives such as solar and wind power are adequate to replace coal power. This poor belief is based on a lack of understanding of the science and realities of energy. We will consider the details in a later chapter.

The natural gas option has some key challenges. But it also has very important advantages. There is no perfect option as is discussed in a later chapter. For several regions, natural gas is an important option in the intermediate term because it offers valuable advantages.

Air Pollution Deaths from Fossil Fuels are Alarming

This piece of misinformation is causing an intense fear and hatred for fossil fuels.

The World Health Organization (WHO) and the Health Effects Institute (HEI)[210] both estimate that air pollution contributes to about 6.5 million global deaths per year[211,212]. The estimate includes deaths from indoor and outdoor air pollution. They attribute 2.5 million deaths to indoor air pollution and 4 million to outdoor air pollution.

What is the role of fossil fuels?

The major source of indoor air pollution is heating and cooking with dirty solid fuels such as biomass[213].

Fossil fuels contribute mainly to outdoor air pollution. But not all outdoor air pollution is from fossil fuels. There are several other sources of outdoor pollution such as industrial activities, burning of waste and agriculture, and wildfires.

What is the specific impact from fossil fuels on outdoor air pollution?

Studies provide a wide range of estimates for the contribution to outdoor air pollution from fossil fuels. It is logical to only consider the estimates that are consistent with WHO and HEI. A recent study, which is a collaboration between several academic institutions, provides such an estimate[214,215]. The study has two key advantages. The total number of deaths from outdoor air pollution in the study is consistent with WHO and HEI data. Moreover, the study provides a breakdown for the contribution from coal, natural gas, and liquid petroleum products.

The study finds that the outdoor air pollution from fossil fuels contributes to 1.1 million deaths per year[216]. Coal is the bigger culprit. It contributes to more than half of the total deaths from fossil fuels, i.e., over 0.5 million deaths per year. Combined, natural gas and liquid petroleum products contribute to 0.5 million deaths per year.

These numbers are in excellent agreement with the most recent Lancet Countdown report[217]. This report is based on the collaboration of scientists and medical researchers from fifty-two research institutions and UN agencies. The report states that there were 1.2 million deaths in the year 2020 from air pollution due to fossil fuels.

Some activists use an outlier study for their numbers. For example, they hype that there are 8.7 million deaths per year because of the outdoor pollution from fossil fuels. Their claim is based on a study that is inconsistent with WHO data, HEI data and the Lancet report[218,219]. The total number reported by WHO and HEI is 4 million deaths per year, which includes all the sources of outdoor air pollution. Apart from fossil fuels, there are many other sources which contribute to these deaths[220]. Fossil fuels contribute to about a fourth of the total deaths from outdoor air pollution based on credible studies[221,222]. Clearly, the study which estimates 8.7 million deaths from fossil fuels alone is not consistent with the global data from credible organizations. The hyped number is eight times higher than the data from credible studies[223,224]. Hyping this inconsistent number is an example of misinformation via the use of outliers.

Every life is precious. Even one additional death is too much. But it is critical to view the deaths in a practical context. We must consider how the ~1 million deaths per year from fossil fuels compare to the global deaths from other causes.

- Use of tobacco contributes to over 8 million deaths per year[225]. This includes 1.2 million deaths that are caused by secondhand smoke. Severe restrictions on the use of tobacco can greatly reduce the deaths.
- Heating and cooking using solid fuels causes over 2 million deaths per year[226,227]. These deaths can be prevented if the solid fuels are replaced by natural gas, liquified petroleum gas (LPG) or other clean sources.
- Obesity causes over 2.5 million deaths per year[228]. Proper diet and physical exercise can radically reduce obesity.
- Traffic accidents cause 1.3 million deaths per year[229]. An expanded use of public transportation and a focus on road safety can drastically decrease these deaths.

Fossil fuels cause slightly above 1 million deaths per year because of air pollution. But they also provide immense benefits to the global society. Specifically, fossil fuels provide over 80% of global energy. Causes such as tobacco and obesity lead to millions of deaths. There are no benefits to society from these causes.

Moreover, causes such as smoking, indoor pollution and obesity are markedly easier to address. Yet, progress has been very slow to decrease the deaths from tobacco, indoor pollution, obesity, and traffic deaths. This informs us about what our society deems as acceptable, i.e., not particularly alarming. We cannot ignore this harsh reality. The inflated alarm about the deaths from fossil fuels is not credible.

Fossil Fuels Receives Trillions of Dollars Per Year in Subsidies

Subsidies are payments or tax credits provided by the government to an individual or organization[230,231]. According to the UN and IEA, fossil fuels on an average receive $500 billion per year in global subsidies[232,233,234]. For reference, $1000 billion = $1 trillion.

Most of these subsidies are provided to the consumers. Many governments provide these subsidies as a relief from energy prices[235].

Activists publicize that fossil fuels receive several trillion dollars in subsidies. The misinformation is increasing anger and frustration.

What is the source of the misinformation? Reports published by the International Monetary Fund (IMF).

A recent IMF report claims that fossil fuels receive $7 trillion of subsidies[236,237]. This value is not consistent with those provided by the UN and IEA.

The IMF report includes the cost of side-effects of fossil fuels such as air pollution and climate change[238]. Their cost estimate for the side-effects is $5.7 trillion. The IMF report includes this value as a subsidy for fossil fuels[239]. The report implies that governments are providing a subsidy by not extracting a price for the side-effects of fossil fuels.

But including a price for side-effects is not consistent with the established definition of a subsidy. Subsidies are payments or tax credits provided by the governments to an individual or organization. Including the costs of side-effects to convey information about subsidies is misleading.

I discuss below why including side effects is not meaningful.

Almost everything (e.g., social media, medicines, surgeries, and technological advances) has side-effects. It is pointless to value the side-effects in isolation. It is well understood that the major advances of modern society are because of fossil fuels.

Any alternative system that could eventually replace fossil fuels will also have substantial side-effects[240]. The cost of the side-effects from the alternative system should also be included for a valid analysis. This report does not include such information. Thus, it tacitly makes two false assumptions.

The first false tacit assumption is that an equivalent alternative to fossil fuel energy is already available. An equivalent alternative is one that has a similar cost and convenience. Most of the commercially available alternatives do not have the same cost and convenience. Moreover, many low-carbon alternatives are yet in the demonstration or prototype stages[241]. The fact is that equivalent alternatives to fossil fuels are not currently available for wide use. For instance, fossil fuels technologies are mature. According to IEA, less than 25% of the technologies required to reduce CO_2 can be classified as mature[242].

The IMF analysis does not include a cost estimate of the side-effects from the alternatives. So, the second false tacit assumption is that an equivalent alternative will not have any side-effects[243].

An objective analysis will never include the cost of side-effects as subsidies. This is because too many assumptions are required to estimate such costs. It is impossible to estimate a meaningful or an objective value for side effects. So, a discussion of subsidies which includes the cost of side-effects is meaningless[244].

For the sake of completeness, I will discuss a few examples of the poor assumptions used in the IMF report.

Historically, the costs from side effects such as air pollution have been mitigated via regulations and other policies. The cost-efficiency for such efforts is very high. For example, the U.S. EPA estimates that the cost of air pollution control technology is thirty times lower than the health costs arising from air pollution[245]. Thus, a $35 billion investment in air pollution control technology can eliminate $1 trillion in health costs. So, the actual cost will be only $35 billion for every $1000 billion that is reported for air pollution. The costs in the report are wildly exaggerated[246,247].

Other assumptions, such as those used to estimate the cost of deaths, also have serious problems. Two assumptions are required for this purpose. One assumption is about the number of deaths. The other is about the societal cost per death. The report uses a high value for the number of deaths. Prior studies provide a wide range of estimates for air pollution deaths from fossil fuels. It is critical to use the estimates from credible studies. The report does not do this. It uses a value for number of deaths that is three times higher than those estimated by credible studies[248,249,250,251,252].

Moreover, it assumes a very high societal cost per death. For example, the report assumes an average cost of $5.2 million per death in a developed country[253]. Such assumptions are abstract. They can be used to provide any cost figure for side-effects that is desired by the estimator.

Finally, let us look at their strangest assumption. The report assumes road congestion, road accidents and road damage as a side-effect of fossil fuels[254]. That is like saying the agriculture industry is responsible for the growing obesity in the global population. This poor assumption has a major impact on the costs. Road congestion, accidents and damage represent the costliest side-effect from gasoline in the report.

The costs discussed in the report are very misleading. Most of the costs do not fall in the common definition of subsidies. Describing the costs of side-effects as subsidies creates a false impression. Moreover, the cost of side-effects from fossil fuels are based on very poor assumptions. The costs are wildly exaggerated. Many activists are not aware of these critical facts.

Chapter Highlights

Activists routinely make false or misleading claims about fossil fuels. Popular examples of misinformation are listed below:

- Firm evidence has been available about the climate impact of fossil fuels for many decades.
- Natural gas has too many problems to be considered as an intermediate term option.
- The air pollution deaths caused by fossil fuels should cause a major alarm.
- Fossil fuels receive several trillions of dollars of subsidies each year.

...§§§-§-§§§...

7. Fossil Energy: Misinformation by Skeptics

Skeptics are misinformed about the relation between fossil fuels and climate change. They do not believe that an unabated use of fossil fuels is a reason for serious concern. So, they see no reason to decrease the use of fossil fuels.

Skeptics have many incorrect views about climate change. I have discussed the highlights earlier. Here, we will review their false beliefs about fossil fuels.

Fossil Fuels are Not Responsible for the CO_2 Rise

Skeptics believe that the rise in CO_2 cannot be attributed to fossil fuels. This is not consistent with the facts. There is a large amount of evidence that shows fossil fuels have been mainly responsible for the rapid rise in CO_2 over the last century.

The burning of fossil fuels produces CO_2. The use of fossil fuels has increased enormously since the industrial revolution. This has resulted in a release of over 2000 billion tons of CO_2[255]. Studies indicate that oceans and plants absorb about half of the CO_2[256]. This is evidenced from an increase in the ocean acidity and fertilization effect on plants. A large fraction of the other half remains in the atmosphere for a long time. This has led to a rise in atmospheric CO_2 levels.

Carbon has three isotopes[257]. These isotopes can be distinguished based on their distinct masses. Fossil fuels have a different distribution of isotopes compared to the other sources of carbon. For example, they do not contain the 14C isotope. The CO_2 from fossil fuels can be estimated because of their unique isotopic footprints[258]. Such studies show that fossil fuels are mainly responsible for the rise in CO_2.

Another piece of evidence is the decrease in atmospheric oxygen content[259]. This is again consistent with the burning of fossil fuels.

Based on the plethora of evidence, scientists are certain that fossil fuels are responsible for most of the CO_2 rise.

Unabated Use of Fossil Fuels Is Not a Serious Concern

This is a popular misbelief amongst skeptics.

Fossil fuels contribute to over 80% of the global energy. Three quarters of the global greenhouse gas emissions are from fossil fuels[260]. The greenhouse gases emissions are changing our climate.

An unabated use of fossil fuels will continue the rapid rise of greenhouse gases. This will cause a large increase in global temperatures.

The impact from climate change is already significant and is expected to worsen with time[261]. The higher the temperature rise, the higher will be the impact on people's lives, and the ecosystems. The impacts of climate change are varied and affect every aspect of our ecosystems. Also, the risk for very high impact, low probability events increase with increasing warming.

Currently, we emit over 50 billion tons of greenhouse gases on a yearly basis[262]. It is critical that we drastically decrease our global greenhouse gas emissions. This will require a major change in the approach to our use of fossil fuels. But it is crucial to make this change carefully. A poorly thought-out plan can cause major energy disruptions. Such a plan can cause immense harm to global society.

Chapter Highlights

Popular examples of misinformation about fossil fuels by the skeptics are listed below:

- The observed warming of the earth is not related to the use of fossil fuels.
- There is no serious concern associated with the unrestricted use of fossil fuels.

...§§§-§-§§§...

Misinformation about Low Carbon Power

8. Low Carbon Power: Misinformation by Activists

The electric power sector is one of the largest emitters of greenhouse gases. It contributes to 40% of the greenhouse gas emissions from the energy sector[263,264,265]. The greenhouse gas emissions from low-carbon power are very small compared to fossil fuel power. Hence, low carbon power is a key tool to mitigate climate change.

The importance of low-carbon power makes it a prime target for misinformation. Here, we will review the misinformation by activists.

Cost of Solar & Wind Power Can be Compared to Fossil Fuel Power

This is a very popular myth. It is based on the incorrect use of a metric called levelized cost of electricity (LCOE).

The LCOE metric accounts for the upfront cost and operating and maintenance costs[266]. But it does not account for the difference *in the nature of electricity production* between the power options. This is a major flaw of this metric. Fossil fuel and nuclear power can provide electricity on demand or when we need it. Solar power can provide electricity only when there is sunlight. Wind power can do so only when the wind is blowing. The LCOE metric does not account for this major deficiency of solar and wind power.

Why is the nature of electricity production so important? Because our society has stringent electricity needs.

We need continuous electricity for certain applications at our homes, industries, and other facilities. For other applications we need electricity in short bursts. The demand for electricity fluctuates with time. Our current lifestyle is possible because the electrical grid can supply electricity 24X7 when we need it.

Unlike fossil fuel power, operators cannot control the output from solar and wind power. Weather controls their output. The U.S. Energy Information Administration (EIA) lists solar and wind power in a separate category because of this important distinction[267]. Solar and wind are listed as resource constrained technologies. Whereas fossil fuel power plants are listed as dispatchable technologies.

Dispatchable technologies can provide electricity when required. They do not suffer from the deficiency of resource constrained technologies. LCOE estimates do not account for this critical difference in technologies. EIA lists solar/wind power and fossil fuel power in separate categories to prevent the direct LCOE comparisons between the technologies.

It is critical to maintain a balance between electricity supply and demand. The deficiency of solar and wind power makes it more difficult to maintain the balance[268,269].

To maintain the balance, grid operators are forced to take certain actions.

When solar and wind are producing electricity at peak levels, there is overproduction. Dispatchable power plants are turned down to compensate. The dispatchable power plants are forced to sacrifice their performance to accommodate solar and wind power.

Some form of back-up is required when there is no sunlight or wind. Based on a global average, sunlight is available for less than seven hours per day[270,271]. Similarly, the limited availability of wind also poses significant problems. The back-up requirements for solar and wind power can be extensive to maintain a balance between supply and demand.

The actions required to maintain the supply and demand balance have a cost. Higher levels of solar and wind power result in higher costs[272,273].

The Organization for Economic Cooperation and Development (OECD) have co-published a study on this topic[274]. The study discusses the costs related to solar and wind power that are not included in LCOE estimations. The extra costs arise because of the deficiencies of solar and wind power.

Let us review the major deficiencies of solar and wind power. The power output is not constant. It is variable because sunlight and wind are not available all the time. There is uncertainty in power generation because the amount of sunlight and wind varies hour-to-hour, day-to-day and season-to-season. The transmission and distribution of electricity is more complex because of the inherent attributes of solar and wind power. These deficiencies add substantial costs.

These costs are not included in the LCOE metric. Activists and the media focus on the LCOE metric for cost comparison. So, the costs that are routinely reported do not include all the costs of solar and wind power. Only partial costs of solar and wind power are reported. The reported costs do not reflect the true costs for solar and wind power. So, such cost comparisons are not meaningful.

I will discuss an example to further illustrate this point.

Consider a small hospital which requires a nurse to monitor its patients overnight. Nurse Jill is available every night for the required number of hours. But nurse Jane is only available for four hours a night. Her availability varies over the year. She is more available in the first half of the year.

The presence of a nurse every night is essential. So, the irregular availability of nurse Jane is a major challenge for the hospital. As compared to nurse Jill, the hospital will need extra resources if they hire nurse Jane. Extra resources will be needed to hire, conduct background checks, and train new nurses and to maintain a complex schedule with more frequent backup needs.

The extra resources are a result of the deficiencies in the availability of nurse Jane. The related costs must be included for a valid comparison. A direct comparison of their hourly rate alone–which does not include the costs from the extra resources–will lead to a poor decision. To directly compare the hourly rate of nurse Jane with nurse Jill is similar to directly compare the LCOE of solar or wind power with fossil fuel power.

Solar & Wind Power Will Provide Cheaper Electricity

Several pathways have been proposed for a decarbonized future, i.e., a net zero world. A massive use of solar and wind power is a common feature in all the pathways.

Recent studies claim that solar and wind power will provide cheaper electricity than fossil fuels in a decarbonized future[275,276]. These studies use several speculative assumptions in their estimation. For example, major assumptions are made about the grid interconnectivity, grid stability, and the future technology costs, and availability of technologies that are still in the demonstration or prototype stages. These assumptions have large uncertainties. Also, the conclusions are very sensitive to these assumptions. Such studies are speculative and have an enormous potential for researcher bias. Most importantly, these studies ignore basic science.

What is basic science? It is undisputable knowledge that does not change with new findings. I use basic science to show that these studies are wrong.

First, some background. For a valid cost comparison of power technologies, their final product must be 24X7 electricity. Solar and wind power cannot provide 24X7 electricity on a standalone basis. But they can do so using extra aids. For example, solar and wind power can provide 24X7 electricity by adding energy storage. Other aids include the overbuilding of solar and wind power and extending the transmission infrastructure. All extra aids are costly. These costs must be included for a valid comparison with fossil fuel power[277].

Solar, wind, and fossil fuel power technologies have two key similarities. From basic science, we know that solar energy is the primary source of energy for solar, wind and fossil fuel power[278]. Also, they must provide the same final product, i.e., 24X7 electricity. The power technologies have the same primary energy source and must deliver the same final product. Thus, they are a part of the same basic energy system.

Which power technology has a lower cost? The one that requires a lower resource intensity to provide 24X7 electricity[279]. The resource intensity of a technology is a measure of the intensity of all the resources it requires to provide the product. Examples of resources include labor, materials, minerals, energy, land, and water.

The power technologies need varied resources such as materials, fuel, land, labor, and water[280]. Solar and wind power require far more materials and land. While fossil fuel power requires very large amounts of fuel. Clearly, the resource intensity for the technologies cannot be compared by simply adding up these resources. That would make no sense.

Only basic science can provide such information. The details are discussed below.

The resource intensity of any system depends on the amount of help received from nature. More help from nature equals a lower resource intensity.

For example, consider a flight from New York to London. Less time and fuel are required for the flight when there is a strong tailwind, i.e., favorable wind direction. This favorable wind direction is help provided by nature. The resource intensity–and thereby the cost–is lower when nature provides a helping hand. So, the technology which receives large help from nature has a lower resource intensity.

Nature provides different amounts of help to the different power technologies. How to determine the relative help received by power technologies?

To do so, we must first consider the basic challenges associated with their primary energy source, i.e., solar energy.

Solar energy has two major challenges from the viewpoint of converting it to 24X7 electricity.

- First, solar energy is extremely dilute[281,282]. While earth receives gigantic amounts of solar energy, the amount of energy received per unit area is small[283]. So, it is challenging to capture solar energy–which is a required step to convert it to electrical energy. It is difficult to harness energy from an extremely diluted energy source. For example, consider the challenges related to fishing in a large lake that only has a few fish.
- Second, solar energy is intermittent. Therefore, providing 24X7 electricity using solar energy is an immense challenge.

A helping hand from nature can lower the level of challenge. The technology that receives higher help from nature will have a lower challenge and thereby a lower resource intensity.

How much help is received by each technology?

Let us first consider fossil fuel power. We know how fossil fuels were formed. The solar energy captured by ancient plants and organisms has been converted to fossil fuels[284]. Nature has enabled this conversion process by applying heat and pressure on the plants and organisms in the earth's crust over millions of years **(Figure 8.1).** Specifically, nature has transformed the dilute, and intermittent solar energy into high-energy-density fossil fuels that are available 24X7 for energy production. Properties such as energy density and power density provide information about how dilute the energy source is. Based on any reasonable metric, solar energy is more than a thousand times dilute compared to fossil fuels[285,286].

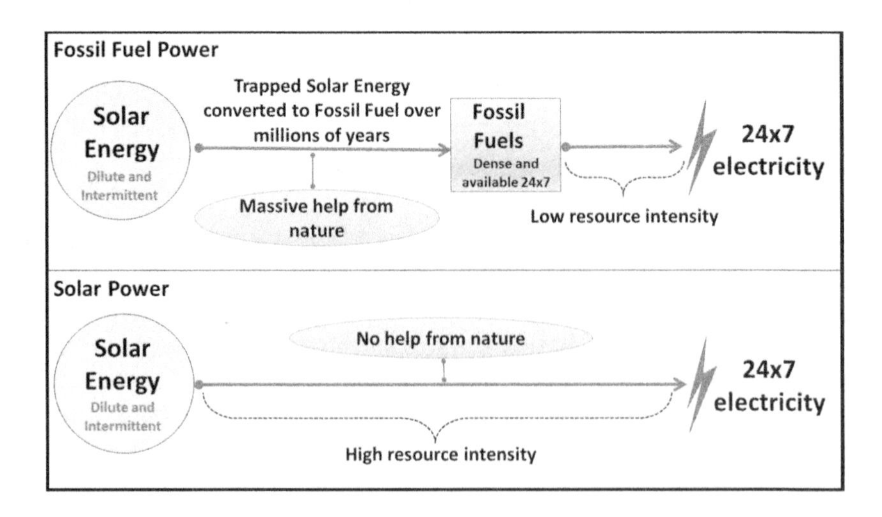

Figure 8.1 The basic science about 24X7 electricity production from fossil fuel, and solar power. It shows why fossil fuel power has a markedly lower resource intensity compared to solar power.

Nature has drastically lowered the resource intensity in the case of fossil fuel technologies by easing the two critical challenges of solar energy. Because of the massive help from nature, fossil fuels are abundant, accessible, easy to transport, store, and convert to usable energy. Thus, nature has enabled an easy path to 24X7 electricity for fossil fuel power.

What about solar power? Solar energy is the direct source of energy for solar power. Nature does not help with the dilute nature and intermittency of solar energy. Thus, solar power has a far more challenging path to 24X7 electricity compared to fossil fuels. It requires a much higher resource intensity. This translates into a markedly higher cost for 24X7 electricity.

What about wind power? From basic science, we know that nature acts upon solar energy to generate wind. But wind is also very dilute and intermittent[287,288]. Nature does not provide a substantial help to wind power. Thus, wind power also has a far more difficult path to 24X7 electricity compared to fossil fuels. It requires a much higher resource intensity. This translates into a markedly higher cost for 24X7 electricity.

A net zero world will require a massive overbuilding of solar and wind power along with massive amounts of long duration energy storage to provide 24X7 electricity[289]. Long duration storage, which can extend from days to weeks, is currently very expensive. For example, the upfront cost of solar power with just 12 hours energy storage is currently five times higher than natural gas power to provide the same amount of annual electricity[290,291,292]. Technologies for long duration energy storage are still in the demonstration stages[293].

Costs of long duration energy storage will decline with technology advances and higher deployment levels. But that will not change the basic conclusion. The cost of solar and wind to produce 24X7 electricity will remain markedly higher than fossil fuel power. Why? Because they have a much higher resource intensity. The costs cannot drop below the level defined by the resource intensity. *The laws of nature dictate this.*

A high resource intensity is intrinsic to solar and wind power because of the lack of help from nature. Basic science dictates a markedly higher cost for electricity in a net zero world, which will be dominated by solar and wind power.

Why is the literature on this topic confusing?

Over the decades, researchers have used many metrics to compare fossil fuel power with solar and wind power. Examples are materials intensity, fuel intensity, land use and energy return on investment (EROI). Fossil fuels are superior based on some metrics, while solar and wind power on others.

Also, some metrics such as EROI require many assumptions because of the unknowns. Consequently, researchers have estimated a wide range of EROI for the different power technologies[294,295]. Some studies have estimated fossil fuel power to have a superior EROI, while others have favored solar and wind power. There is confusion in literature even when considering an isolated metric such as EROI.

Literature has focused on isolated metrics and speculative assumptions. They have not used a basic science approach to compare the technologies. This is a major flaw. A holistic comparison is not possible by focusing on different isolated metrics. A comparison of total resource intensity, i.e., intensity of all the required resources, is crucial for holistic answers.

Only basic science can be used to compare the total resource intensities of the different technologies. The power technologies require a varied nature of resources such as materials, fuel, land, labor, and water[296]. So, the resource intensity for a technology cannot be estimated by simply adding up the resources. A basic science approach is the key to compare the resource intensities and costs between fossil power and solar and wind power.

The quote from Albert Einstein is particularly relevant[297]. "*Look deep into nature, and then you will understand everything better.*"

Residential Solar Power is a Low-Cost Option

Solar panels can be used in small residential applications (residential solar) as well as large utility-scale power plants. Residential solar is also referred to as roof top solar. It offers several benefits. For example, it produces electricity on-site and does not need extra land. Activists believe that residential solar is a low-cost option for producing low carbon electricity. That is not true because of its high cost of electricity production.

The cost of electricity generation from residential solar is two to three times higher than from utility-scale solar in most regions[298,299,300,301]. Representative data from the United States is shown in **Figure 8.2**[302]. The overall benefits offered by residential solar cannot compensate for its massive cost disadvantage in producing electricity[303,304,305].

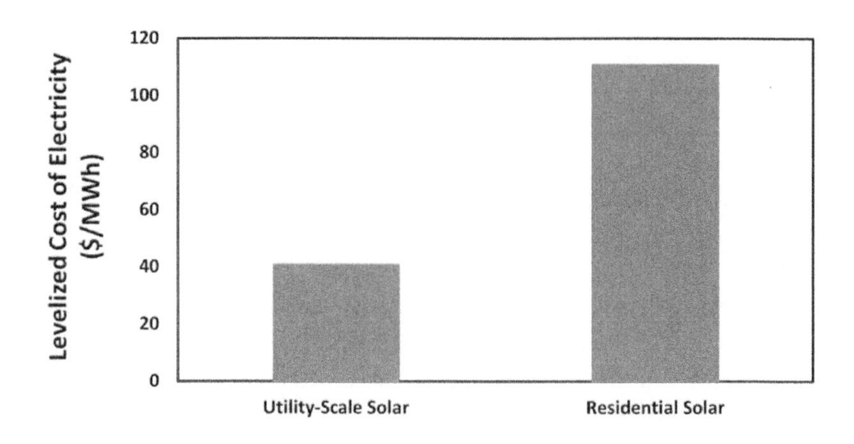

Figure 8.2 Total lifetime cost comparison between utility solar and residential solar to produce electricity in the United States. Data Source[306]: NREL

Residential solar has a markedly higher cost because of its tiny scale. Residential solar systems produce very small amounts of electricity. In contrast, utility-scale power plants produce electricity on a very large scale. Consequently, a utility-scale power plant has an advantage of economy of scale. Economy of scale refers to the principle that a large-scale production often has a lower cost per unit of the product as compared to a small-scale production[307].

The overhead costs are much higher for small-scale applications. A typical home has 15 to 20 solar panels. Each home has a cost associated with system design, permitting, inspection, and customer acquisition. Utility-scale power plants have millions of solar panels. The overhead cost is markedly lower for utility-scale power plants because it allows a far more efficient use of resources.

Residential solar coupled with energy storage is extremely costly[308]. This system involves two technologies deployed at a very small size. Such a system suffers from a huge economy of scale disadvantage. This leads to a very high cost.

Why is there a confusion about the costs of residential solar?

Large subsidies are available for using residential solar in many regions[309]. A technology that receives a large subsidy can be wrongly perceived to have a low cost. This has led to the false notion that residential solar is a low-cost option.

Solar & Wind Power Have a Large Share of Global Electricity

Media often touts that the use of solar and wind power has increased massively. This has led to confusion about the share of solar and wind power. Activists believe that solar and wind power have a large share in global electricity production.

That is not true. The share of solar power is 5% and that of wind power is 8%[310,311,312].

What is the share of all the low carbon power technologies? In 2000, the combined contribution of low carbon technologies to the global electricity was 36%[313,314,315]. This contribution only increased to 39% in 2022[316].

Why was the increase so modest? There are three reasons for this. First, the use of fossil fuel power increased markedly over this period **(Figure 8.3)**[317]. Second, the use of nuclear power did not increase. Third, the increase in solar and wind power was a small fraction of the total increase in electricity demand. The electricity generation from solar and wind power increased by 3500 billion kWh[318]. But the total electricity demand increased by 14,000 billion kWh.

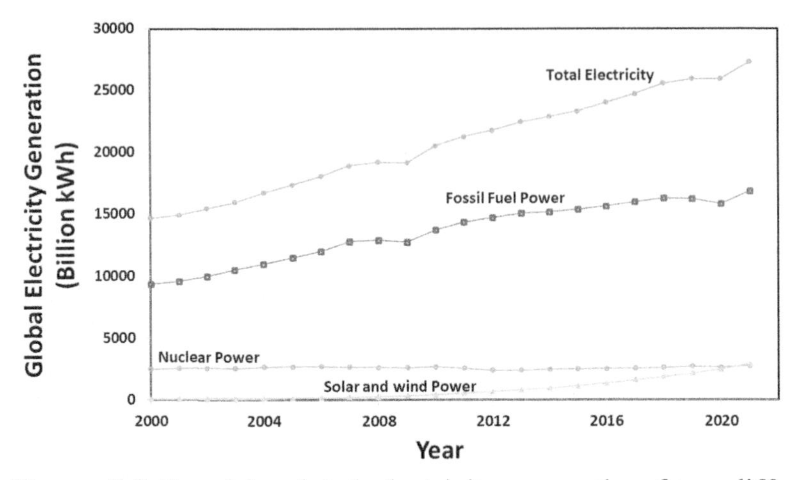

Figure 8.3 Trend in global electricity generation from different sources by year. Data Source[319]: U.S. EIA

Solar and wind power play a much larger role in certain countries. For example, the share of wind power in Denmark is 45%[320]. Activists hype that Denmark is an example that others should follow. But this example is not applicable to most countries.

The high percentage of wind power is possible in Denmark because of its unique situation. Neighboring countries assist with the supply and demand imbalance caused by its large share of wind power[321]. The excess electricity produced in Denmark–when the wind speeds are high–is absorbed by Sweden and Norway. These countries provide electricity to Denmark when wind speeds are low. So, the neighboring countries serve as an energy storage system for Denmark's wind power[322].

Denmark has a low electricity consumption[323]. So, it needs to import or export only a small amount of electricity from its neighbors to maintain a balance. Also, its electrical grids are well connected to its neighbors. This allows for a smooth transfer of electricity when required.

Denmark would have to curtail electricity production and face blackouts in absence of these unique conditions[324]. These unique conditions do not apply to most of the global electricity production. Thus, the hype about Denmark's large share of wind power is misleading.

Intermittency of Solar and Wind Power is Not a Major Issue

Electricity produced from solar and wind power is intermittent. The amount of electricity produced varies over hours, days, months, seasons, and years. Activists do not believe that this is a major challenge for a shift to solar and wind power. As discussed below, this belief is wrong.

The current energy storage capacity is trivial[325]. So, electricity is produced at the instant when there is demand. The demand for electricity varies with time. The electricity demand depends on the region, weather and level of equipment use[326]. **Figure 8.4** shows an example of the electricity demand over a 24-hour period in the United States[327]. The demand is typically at its lowest before we wake up. The peak demand is at some point when most of us are awake. Often peak demand is observed after people return home from work.

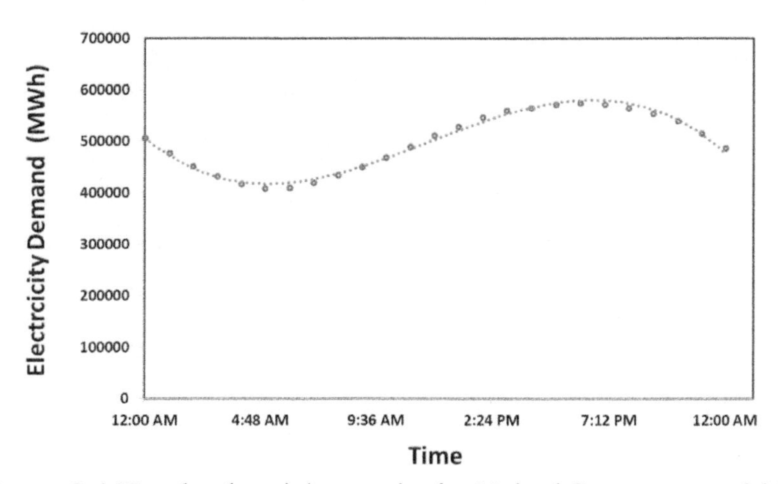

Figure 8.4 Hourly electricity use in the United States over a 24-hour period. This specific data is from June 21, 2023. Data Source[328]: U.S. EIA

The electricity demand needs to be exactly met at each point in time. This is less challenging with fossil fuel power. The electricity output from fossil fuel power plants does not depend on weather. The power plant operator can control the output from fossil fuel plants[329].

The electricity produced from solar and wind power is not uniform or consistent. Solar and wind power produce high levels of electricity during certain periods and low levels in other periods. The power plant operator has no control over this[330]. So, meeting the electricity demand is a challenge.

Many regions use solar and wind power. What is the impact on the electrical grid system?

The drawbacks of solar and wind power impose a cost on the system. The flexibility of the electrical grid system determines how much solar and wind power can be used without a large increase in the cost.

Every system has a certain level of flexibility to ensure year-round robust operations. Flexibility is the ability to handle a higher or lower electricity demand than normal. This flexibility allows most grid systems to easily accommodate low levels of solar and wind power.

The level of flexibility depends on the features of the system.

A system with a large share of natural gas power is more flexible. Natural gas power plants provide flexibility because they can be adjusted up or down to meet the electricity demand[331,332].

A system is also more flexible if its neighboring grids can receive and provide a significant amount of electricity when needed.

Some grid systems are more flexible than others[333]. But each system only has a limited level of flexibility because of economic reasons. A grid system with a higher flexibility can accommodate more solar and wind power.

The cost to the system for accommodating solar and wind power increases with an increasing share of solar and wind power[334,335].

The California system is a good example to discuss because it has a substantial share of solar power[336,337].

Solar power produces peak electricity during the sunny middle hours of the day. The California system ramps down natural gas power plants to balance some of the high electricity output from solar[338,339]. It also exports some electricity to the neighboring grids during this period. After the system has used up its flexibility, it is forced to curtail some electricity from solar. To curtail means to stop production, which is the same as wasting electricity.

The situation reverses in the evening. Solar power produces low levels of electricity in the evening when there is a peak demand. The California system ramps up its natural gas power plants in response. But this is not enough to meet the demand. So, it has to import electricity to meet the peak demand in the evening hours.

Such actions increase the system cost. Let us consider the two obvious costs first.

The curtailment of solar power is an obvious cost. Curtailment has an economic and environmental cost because of the energy waste. The extent of curtailment depends on the share of solar power.

The share of solar power in the total electricity production was low in California a few years back[340]. The California system could use its system flexibility to minimize curtailment at those low levels of solar power. It was possible for the system to adapt to the excess electricity from solar. Little to no curtailment of solar power was required then.

But more and more curtailment has been required over the years because of the increase in the use of solar power[341]. The available flexibility in the California system is not adequate. Further ramping down of natural gas power plants is not practical. Further export is also not possible because of grid limitations[342]. So, the increase of solar power in California has caused a marked increase in curtailment over the years. The curtailment is especially severe in the spring season. Solutions to decrease the curtailment are being explored, but they are not cheap.

The California system curtailed over 2 billion kWh of electricity from solar power in 2022[343]. How much is this curtailment? Over 65 countries in the world use less electricity than this per year[344]. The African nation Chad, which has a population of 18 million, uses only a tenth of the electricity curtailed by California[345].

The California grid system has a decent level of flexibility. It has a large fleet of flexible power plants[346]. It also has reasonable access to neighboring grids to export and import electricity. But this flexibility is not adequate. Solar and wind power produces only 25% total electricity in California. Yet even this share of solar and wind power has a large negative impact. It forces a substantial amount of curtailment.

The large electricity imports that are required are another obvious cost to the California system. California imports a significant amount of the electricity to meet the period of peak demand[347,348]. Recall, electricity production from solar is very low during this period. This drawback of solar is costly because the price of electricity is high during the periods of peak demand.

There are two less obvious costs to the grid system because of the deficits of solar and wind power. We will consider these next.

The first cost is incurred because dispatchable technologies are forced to operate in a suboptimal manner. Dispatchable plants such as natural gas power plants are forced to pay the price for the deficits of solar and wind power. Recall, the system is forced to ramp down its natural gas power plants to accommodate the high production of solar power[349]. The forced lower production rates hurt the efficiency and economics of the dispatchable power plants.

The result is a neglect of these power plants and related facilities. Less money is spent to maintain their reliability. An inadequate spend on reliability can have a major impact when the system is under stress during extreme weather events[350]. The recent blackouts in Texas during cold weather serve as a representative example.

The forced poor economics can also lead to the premature shutdown of dispatchable power plants[351]. This can cause more problems for grid reliability. Thus, the overall efficiency of the system and its robustness decreases with a higher use of solar and wind power. Yet, this impact is seldom discussed.

The second cost that is less obvious is related to the loss in flexibility of the grid system. A grid system which has a high level of flexibility is more robust during periods of heavy stress. A grid system can be exposed to stress because of extreme weather, failure in transmission lines, and unplanned shutdown of power plants[352]. This occurs occasionally in most grid systems. We saw earlier that the flexibility of the system decreases because of solar and wind power. *The available flexibility in the system decreases as the share of solar and wind power increases.* This exposes the system to a greater risk of blackouts during the periods of stress.

The intermittency and uncertainty associated with solar and wind power causes a mismatch between supply and demand in the grid system. The system can accommodate the deficiencies of solar and wind. But there is an associated cost. The cost increases with an increasing share of solar and wind power in the grid system[353,354]. Most grid systems have a comparable or lower flexibility compared to the California grid system. So, most regions cannot accommodate over 25% of solar and wind power without markedly increasing the cost and other negative impacts[355]. For most regions, a reasonable share of solar and wind power that is available at a low cost will lie between 10% and 30%. As more grids increase their share of solar and wind power, their capacity to import and export electricity will decrease[356]. This will decrease flexibility in the grid systems and make it more costly to increase the share of solar and wind power.

Above, we looked at the impact of the intermittency of solar and wind power on the current grid systems. Next, we will consider the impact on a net zero world.

The electricity demand will have to be met 24X7 each year even in a net zero world. There will be many challenges to meet this goal. But there is one key challenge. Even large regions can experience a few 24+ hour events of cloud cover and low wind speeds (aka dark doldrums) each year[357,358]. The electricity produced from solar and wind power will be drastically reduced during such periods for the entire region. A very high share of solar and wind will make it very challenging to meet the electricity demand during these periods. So, expensive long duration energy storage or back up will be required even for the grids that are interconnected over large regions.

Why is this not a problem currently? Because the share of solar and wind power is low and the share of dispatchable units is high. The situation can be handled by ramping up dispatchable units.

But the situation will be different in a net zero world. The share of solar and wind power will be very high because they will be the major producers of electricity[359,360]. A ramp up of a small fleet of dispatchable units will be grossly inadequate to meet the electricity demand during periods of low sunshine and wind[361].

There is one more major challenge in the net zero world. The electricity demand will be much higher because of massive electrification[362]. According to IEA, it will be factor of two to three times higher than today[363]. A practical way to meet the electricity demand during extended periods of low sunshine and wind will be via long duration energy storage[364]. A large amount of electricity will need to be stored in advance to meet the demand during the low output periods. This is true even for continent sized grid interconnectivity[365,366,367]. The long duration storage, overbuild of solar and wind power, and massive grid expansion that will be required will be expensive.

Fossil fuel power technologies have a large inherent advantage because of the enormous help from nature[368]. This allows fossil fuel power to easily meet 24X7 electricity demand. Solar and wind power do not have such help from nature. Basic laws of science and nature, thus, dictate that the electricity costs will be much higher for solar and wind power. Nothing can change this.

Consider an example from the world of arts. Nature provides a gift of exceptional creativity to some artists. For example, consider the geniuses Mozart and Picasso. An average person–who does not have any special gift from nature–can never make comparable contributions, irrespective of the amount of hard work. A massive help from nature is extremely valuable and impossible to beat.

Activists ignore such basics. They use poor assumptions to argue that the shift to solar and wind power will lower electricity costs. They do not consider the massive help provided by nature to fossil fuel power compared to solar and wind power.

Many are excited about options such as energy storage, microgrids, virtual power plants, massively expanded grids, massive overbuilding of solar and wind power and vehicle to grid technology[369]. But all the options add substantial cost and increase the system complexity.

The costs will be markedly higher in a net zero world, even in the most optimized scenario. This is true for any option or combination of options. Why? Because the basic laws of science and nature must be obeyed.

The quote from Richard Feynman is an excellent reminder.

"For a successful technology, reality must take precedence over public relations, for nature cannot be fooled."

Nuclear Power is Too Dangerous to be An Option

Activists believe that nuclear power cannot be used safely. This is not true based on the historical records. Nuclear power has been safely used for a long time. The first commercial plant was deployed in the 1950s. It is second only to hydropower in terms of low-carbon electricity generation. Since the 1980s, it has contributed to roughly 15% of the global electricity[370].

The contribution has been higher in certain western countries. Nuclear power has provided 20% of the electricity in the United States for three decades[371,372]. The share has been even greater in some other countries. For example, nuclear power has provided 70% of the total electricity in France since several decades. This would not have been possible if nuclear power was dangerous, or the projects were extremely challenged.

How safe are nuclear power plants? Studies have compared the deaths related to different power technologies over time[373,374,375]. For a fair comparison, the data must be expressed in terms of the number of deaths per unit of electricity produced. The deaths related to nuclear power were comparable to solar and wind power[376]. Deaths related to coal power plants were several hundred times higher.

Nuclear power has caused three high-profile accidents[377]. Activists use these to claim that nuclear power is too dangerous.

The example of air travel is useful to expose the fallacy of the argument. Every so often there are major airplane accidents. We have not limited or stopped air travel because of these accidents. We consider air travel to be safe despite these accidents. This is because the overall safety record of air travel is excellent. Similarly, a few accidents do not make nuclear power unsafe. *We must consider its overall health and safety record, which is excellent.* Moreover, the learnings from the prior accidents are being used to further improve the safety of the process[378].

Nuclear power has some key challenges. Projects are complex, expensive and can require a long time to complete. Also, disposal of radioactive waste from the nuclear plants can be an issue. But these challenges can be managed. This is especially true if a diverse energy mix is used, and projects are planned carefully and strategically. The substantial use of nuclear power around the globe for several decades is evidence of this.

Moreover, nuclear power has critical advantages over solar and wind power. It does not depend on the weather and can provide 24X7 electricity. Nuclear power plants require less land and have a much longer life. Nuclear power is an important option because of its important advantages. I will discuss the crucial importance of a diverse energy mix in a later chapter.

Chapter Highlights

Activists routinely make false or misleading claims about low carbon power. Popular examples of misinformation are listed below:

- The cost of solar and wind power can be compared with fossil fuel power via the commonly used metric of levelized cost of electricity (LCOE).
- Solar and wind power will provide cheaper electricity than fossil fuel power in a net zero world.
- Residential solar power is a cost-effective option for reducing greenhouse gases.
- Solar and wind power already play a major role in global electricity production.
- The inadequacies of solar and wind power are not a major issue for the electrical grid.
- Nuclear power is too dangerous to be considered as an option to reduce greenhouse gases.

…§§§-§-§§§…

9. Low Carbon Power: Misinformation by Skeptics

Most skeptics have a low opinion of technologies such as solar and wind power. They do not believe that climate change is a serious problem. So, they see no reason to switch away from fossil fuel power.

Their firm beliefs have led to key misconceptions. We will examine these below.

Greenhouse Gas Emissions from Solar and Wind are Not Low

Emissions must be compared on a lifecycle basis. A lifecycle extends from the cradle to the grave. The life cycle of a power plant includes extraction of raw materials, production of components, installation, power generation, maintenance, operation, decommissioning, and disposal.

A substantial amount of energy is required upfront for solar and wind power. For example, energy is required to prepare the land, extract minerals and produce materials. Solar and wind power require several times more land, minerals, and materials compared to fossil fuel power[379,380]. So, the upfront energy requirements are markedly higher for solar and wind power. Currently, fossil fuels provide this energy. Based on this, skeptics believe greenhouse gas emissions are not that low for solar and wind power. As discussed below, this is not true.

Solar and wind power generate electricity for twenty-five years. During this period, they do not emit greenhouse gases because they do not use fossil fuels to generate electricity[381]. The greenhouse gases emitted by solar and wind power from the upfront energy use are small compared to the ongoing emissions from fossil fuel power plants. Studies have compared the life cycle greenhouse gas emissions from the power technologies **(Figure 9.1)**[382,383,384,385]. These studies show that solar and wind power emit far lower greenhouse gases.

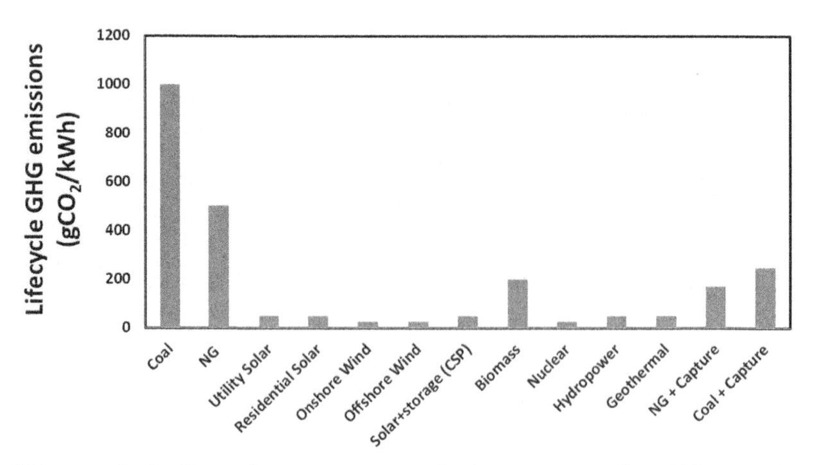

Figure 9.1 Greenhouse gas emissions over the life cycle of various power technologies. Data Source[386,387]: UNECE, NREL.

Solar and Wind Power are Not Reliable Options

Solar and wind power is markedly more challenged than fossil fuel power to provide 24X7 electricity. But this does not mean that solar and wind power cannot be used reliably.

Several options are available to use solar and wind power reliably at low-to-moderate deployment levels. Options include the adjusting of electricity production from dispatchable power plants and transferring electricity to and from the neighboring grids.

Many countries have a history of reliable operation of electrical grids with a 10% or more share of solar and wind power[388,389]. These countries would have stopped the use of solar and wind power if they caused major reliability issues. But that has not been the case.

Solar and wind power can also be used reliably at much higher levels using aids such energy storage and extended grids. Solar and wind power have very low lifecycle greenhouse gas emissions. Therefore, solar and wind are important options to decrease emissions from the power sector. But it is critical to understand the limitations of solar and wind power and the cost implications. Related issues were discussed in an earlier chapter.

Chapter Highlights

Skeptics often share information about low carbon power which does not align with facts. Common examples of misinformation are listed below:

- Solar and wind power do not have low greenhouse gas emissions.
- Solar and wind power have too many problems to be considered as options for electricity production.

…§§§-§-§§§…

Misinformation about Electric Vehicles

10. Electric Vehicles: Misinformation by Activists

Conventional vehicles dominate the transportation sector. Fossil fuels such as gasoline, jet fuel and diesel are used in conventional vehicles.

The transportation sector contributes to 16% of the total greenhouse gas emissions[390,391,392]. Cars, freight trucks, aircraft, and ships are the major contributors[393,394]. Here, the term "cars" is used to include sport utility vehicles, crossovers, cars of all sizes, light trucks, and minivans[395].

Several options are being evaluated to decrease the greenhouse gas emissions from the transportation sector. The most popular option is battery electric vehicles[396]. I discuss these as electric vehicles.

Electric vehicles have important advantages. They are far more energy efficient than conventional vehicles. They can markedly decrease greenhouse gases and air pollution from the transportation sector[397].

But electric vehicles also have key challenges. These challenges are mostly disregarded by activists. This has led to misinformation about this topic.

Electric Vehicles are Already Having a Substantial Impact

The use of electric vehicles has increased in recent years. Activists tout that this increase has led to an impressive reduction in greenhouse gases. That is not true. The impact from electric vehicles is very small currently.

IEA reported that the fleet of electric vehicles reduced the global oil use by 0.7 million barrels per day in 2022[398]. For reference, the global oil use is 100 million barrels per day[399].

What does this mean in terms of greenhouse gas reduction? In 2022, the global fleet of electric vehicles reduced the global greenhouse gas emissions by 0.1%[400]. This is similar to replacing just 1% percent of the global coal power with natural gas power[401].

A Global Switch to Electric Cars will have a Massive Impact

There are over 1.2 billion cars worldwide[402]. The current share of electric cars is less than 3%[403]. Electric cars are very popular amongst activists. They believe that a 100% switch to electric cars will cause a dramatic decrease in greenhouse gas emissions. This is a false belief.

The transportation sector contributes to 16% of the greenhouse gas emissions[404,405]. Cars are only a part of this sector. The cars subsector contributes to 7% of the emissions[406].

This informs us that electric cars at a maximum can reduce the global greenhouse gases by only 7%. This maximum is only possible if the conventional cars are replaced by electric cars that use low carbon electricity.

That is not the case today. Many countries do not use low carbon electricity[407]. Some countries such as Canada and France do. But countries such as China and India use high carbon electricity. Others such as the United States and Germany are in the middle. Electric cars emit 50% fewer greenhouse gases compared to conventional cars when considering the global average carbon intensity[408,409]. So, electric cars can reduce greenhouse gases by 50% from the global cars sector.

We can use the above information to estimate the total emission reduction for an overnight switch to 100% electric cars. The global greenhouse gases would only reduce by 3.5%[410].

A Wide Adoption of Electric Cars Will Be Easy

This is a common belief held by activists. It is not true because electric cars do not satisfy certain attributes. A wide adoption requires low cost and convenience. Electric cars face critical challenges in these areas.

Electric cars have a high upfront cost or retail price. Claims that the retail price of electric cars is comparable to conventional cars are false. They are based on misleading comparisons. The only robust approach is to compare car technologies with the same model and trim[411]. The retail price of electric cars is 40% to 60% higher than conventional cars based on a robust comparison[412,413]. For example, the retail price for 2023 Kona Electric (SE) is 52% higher than the 2023 Kona (SE).

How do electric cars compare with hybrid cars[414]? Electric cars have a 30% to 35% higher retail price than hybrid cars[415,416].

Also, the lifetime cost of owning electric cars is not low in most regions. Lifetime cost includes all costs over the life of the car. Examples of costs are retail price, financing, fuel or electricity, maintenance, and insurance. The lifetime cost of electric cars is higher than conventional cars, or hybrid cars in many regions[417,418,419,420]. The lower operating and maintenance cost of electric cars cannot compensate for their higher retail price in these regions[421].

Electric cars are also far less convenient. This is a key challenge for the wide adoption of electric cars.

The travel range of electric cars is lower than conventional or hybrid cars. Travel range is the total distance that can be travelled on a full fuel tank or electric charge. The travel range of a conventional or hybrid car is over 50% higher than a typical electric car[422]. The cost of electric cars is markedly higher for a higher travel range[423]. Major advances in electric car technology will be required to achieve a travel range similar to conventional vehicles at a comparable cost.

The charging of electric cars can be a major inconvenience. Residential or at-home charging is convenient. But a majority of the global population will have to rely on public chargers because they lack convenient access to residential charging. This is true even for advanced economies. A major fraction in the United States lacks access to convenient charging[424].

Detached homes are well suited for residential charging. United States has a high proportion of detached homes. Yet, a major fraction of its population lacks convenient access to residential charging. The situation is much worse in other countries because they have far fewer detached homes.

Public charging is also important for people who have convenient access. They need public chargers when travelling long distances or when the car is not charged at the residence.

Several studies have noted the importance of public chargers. A recent study reports that the United States will require 2.4 million public and work-place chargers for 26 million electric vehicles[425]. For reference, the United States currently has less than 0.3 million public and work-place chargers[426,427]. A phenomenal increase in the number of chargers will be required if the 250 million conventional cars in the United States were to be replaced by electric cars[428].

The charging speed of electric cars at the public chargers is a major concern. There are two types of public chargers. The slower chargers require several hours to provide a 250-mile travel range[429]. Majority of the public chargers in Europe and United States belong to this category[430,431]. Fast chargers require between 15 to 60 minutes[432]. Gas stations require only a few minutes to provide a 400-mile travel range for a conventional or hybrid car. So, even fast chargers are less convenient. Also, the extra wait time can result in a significant productivity loss.

Early adopters are more likely to accept the inconvenience and high upfront costs. Early adopters are the small percent, 15% of the population, that is eager to try new technologies[433]. Most consumers are less likely to be excited about electric cars until the major problems are resolved.

Some believe that the problems will be resolved quickly. That is not likely.

Batteries are the major contributors to the cost of an electric car. The cost of battery technology has decreased rapidly over the last couple of decades. This has lowered the cost of electric cars. Yet, electric cars are expensive currently. A further rapid decline in battery costs is key to a quick adoption of electric cars.

But the rate at which the battery costs will decline in the future is likely to slow down. Why? Because the factors that caused the previous cost decline will have a smaller impact.

Most energy and chemical technologies have access to low hanging fruit in the early phases of their deployment. These technologies see a large cost decrease in the early phases. But once the low hanging fruit is taken the cost declines are slow. This is also true for battery technology.

The low hanging fruit has been mostly taken in case of battery technology.

A rapid increase in production levels and R&D advances have led to the cost decline in battery technology. An increase in production levels can decrease the cost of a technology via an economy of scale advantage. From history, we know that the economy of scale advantage diminishes after a certain threshold of production[434]. The scale of battery production has already reached a substantial level[435]. Thus, the impact of further increases in the production level of batteries will be lower.

Similarly, the highest impact R&D advances occur in the early decades. Large efforts in battery R&D have been underway for several decades now. The battery technology has already benefited from several high impact R&D advances.

When any technology matures, the impact from economy of scale and R&D becomes incremental. Once this happens, there is no low hanging fruit. The battery technology is now reasonably mature. So, the low hanging fruit is almost gone for battery technology. This means that the cost of electric cars will not continue to decrease rapidly.

Public chargers are currently dominated by the slow type of chargers that require several hours for a 250-mile charge[436,437]. Fast chargers are relatively scarce. Moreover, even the fast chargers require much more wait time than conventional fueling. Also, charging an electric car at a fast charger can be two to three times more expensive than charging at home[438,439,440,441]. Major advances will be required in the charging technology for increasing the charging speed and concurrently reducing costs.

In summary, massive advances will be required in charging technology and batteries to ease the cost and convenience challenges. Such advances will be slow. This is a major challenge for the rapid, wide-scale adoption of electric cars.

EVs Are a Good Option for Freight Trucks, Ships & Aircraft

Freight trucks, ships, ands aircraft carry very large loads. They also travel long distances[442]. The fossil fuels used in conventional vehicles have high energy densities. This allows them to carry large loads and travel long distances.

In contrast, the batteries used in electric vehicles have very low energy densities[443]. A Li-ion battery holds fifty times less energy than gasoline per kilogram.

This is a major challenge for their use in freight trucks, ships, and aircraft. Battery costs for such applications are very high because of poor energy densities. To ensure a reasonable cost, electric vehicles are forced to have a shorter travel range. For example, the travel range of an electric freight truck is a fourth of a conventional freight truck[444].

Also, electric vehicles are much heavier because of their poor energy densities[445]. Vehicle weight is a critical property for freight trucks, ships, and aircraft. The heavy batteries decrease their payload carrying ability relative to conventional vehicles[446]. Thus, electric vehicles require a compromise in travel range and/or payload. The negative economic implications can be substantial. Recharging electric vehicles is also more problematic than conventional fueling. It takes far more wait times. This also requires some compromise.

The current focus of electric vehicles is on applications that require a short travel range. Examples of such applications are freight trucks that travel within a small region, or aircraft with very short routes.

Electric vehicles are not suited for freight trucks that travel long distances and for most ships and aircraft. Massive advances will be required in battery technology before they become suitable.

Chapter Highlights

Activists are too optimistic about electric vehicles. They often share misleading information about electric cars. Popular examples of misinformation are listed below:

- Electric vehicles are already having a large impact on reducing greenhouse gases.
- A global shift from conventional to electric cars will have a massive impact.
- A major shift to electric cars worldwide will be easy.
- Electric vehicles are a good option for long distance freight trucks, ships and aircraft.

…§§§-§-§§§…

11. Electric Vehicles: Misinformation by Skeptics

Skeptics do not see any value in switching to electric vehicles. They see no reason to move away from conventional vehicles. This has led to some misconceptions.

Electric Vehicles Do Not Reduce Greenhouse Gas Emissions

Electric vehicles require large batteries. Energy is required to extract and process the materials for the batteries. Fossil fuels are used to provide most of this energy. The electricity stored in the batteries power electric vehicles. Fossil fuels are the major source of electricity in most regions. Skeptics believe electric vehicles do not reduce greenhouse gas emissions because they indirectly use large quantities of fossil fuels[447]. But this belief does not hold true for most regions.

All the emissions over the life cycle of the electric and conventional vehicles must be considered in the analysis. This includes the emissions from a) the manufacture of vehicles, b) electricity or fuel production and c) vehicle use and disposal. Life cycle studies consider all the factors.

Life cycle studies show that electric vehicles emit lower greenhouse gases in most regions[448,449,450]. The level of improvement over conventional vehicles depends on the carbon intensity of the electrical grid in each region. Carbon intensity of the grid is the amount of greenhouse gases emitted per unit of electricity produced.

Each region requires a separate analysis. This is because each region uses a distinct set of energy sources to produce electricity. The share of the energy sources determines the carbon intensity of the electrical grid.

The electrical grids in regions with a high share of renewable or nuclear power have a low carbon intensity. Electric vehicles emit markedly lower greenhouse gases compared to conventional vehicles in such regions.

Regions with a large share of coal have a grid with high carbon intensity. Electric vehicles have a much smaller positive impact in such regions.

The life cycle data reported by the International Energy Agency is instructive[451,452]. Electric cars can reduce greenhouse gas emissions by over 70% in regions with a low carbon intensity. But they reduce emissions by less than 15% in regions with a high carbon intensity[453]. What about regions with a global average carbon intensity? Electric cars emit 50% fewer greenhouse gases compared to conventional vehicles in such regions[454,455].

Coal often provides the marginal electricity to power the electric vehicles in regions that use large amounts of coal[456]. Yet, even in such regions electric vehicles can reduce greenhouse gas emissions. But the reduction is very low in such regions[457].

The positive impact of electric vehicles on greenhouse gas emissions varies drastically from region to region. This has important implications for energy policies.

Batteries in Electric Cars Need to be Replaced in a Few Years

The batteries used in electric cars lose some capacity to store electricity with time. The rate of loss depends on several factors. But in most cases the rate of loss is slow.

The interest in electric cars is quite recent. So, long-term data on battery performance is not available. The few studies that are available on this topic indicate that the battery life or capacity is not a major issue[458,459]. A battery capacity of 70% is acceptable at end of life[460,461]. Studies have found that the batteries should be able to maintain a capacity of greater than 70% for many years.

A robust approach to understand the battery life is via the warranty provided by the auto makers. Most provide an 8-year, or 100,000 miles guarantee for the batteries[462]. The performance of a product usually exceeds its guarantee. This suggests that the global auto makers are confident that the battery life will be greater than 8 years. The expected range of battery life is between 8 to 20 years[463,464].

Chapter Highlights

Skeptics are too pessimistic about electric vehicles. They routinely share misconceptions about electric cars. Popular examples of misinformation are listed below:

- Electric vehicles do not reduce greenhouse gas emissions.
- The batteries used in electric vehicles need to be replaced every few years.

…§§§-§-§§§…

Misinformation about the Low Carbon Energy Transition

12. Energy Transition: Misinformation by Activists

The energy sector contributes to 75% of the global greenhouse gas emissions[465,466]. A transition to low carbon energy is critical to the net zero emissions goal. Net zero refers to a balance between emissions of greenhouse gases and their removal from the atmosphere[467].

The transition involves three major tasks: a) a shift to low carbon electricity, b) electrification which is replacing the technologies that use fossil fuels with those that use electricity and c) decarbonization of the difficult to electrify applications.

The energy transition will affect every aspect of our life. This topic is extremely complex because of its vast size and scope. These factors make it a prime target for misinformation.

Here, we will explore the popular myths that are being advocated by activists.

We have the Technologies Needed for Net Zero Emissions

Activists claim that all the technologies we need for the transition to low carbon energy are available. This is a very misleading claim.

The energy transition requires a major shift to new technologies. The new technologies must satisfy certain criteria before they are practically available for wide use. Each new technology must at least have a comparable performance and cost to the technology it is replacing. This means that the new technology a) must not have significant performance or convenience issues and b) must not be substantially more expensive.

Access to adequate data about the new technology is critical to minimize the risk from a major shift. Such data is only available after a technology has been in substantial use for several years. Thus, a new technology is only available for a major shift if it has satisfied two basic criteria.

- It must have been in use for several years with a substantial global market share.
- It must have a comparable performance and cost to the conventional technology.

Any given technology is available for wide use only if it meets these crucial requirements.

The new technologies proposed for several key applications do not satisfy these basic criteria. A recent IEA report states that the technologies that will be required to reduce 35% of the CO_2 emissions in a net zero world are yet in the demonstration or prototype stages[468]. These technologies are not even available on the market, let alone be competitive.

I will discuss key technology gaps in the power, transportation, and industry sectors.

Most proposed net zero pathways include a very high share of solar and wind power[469,470]. Every year there are periods with a low access to both sunlight and wind[471,472]. Such periods can extend from many hours to several days even in case of large extended grids[473]. During such periods, solar and wind power can only meet a small fraction of the electricity demand. Supporting technologies will be critical during such periods in a net zero world. Long duration energy storage technologies such as batteries and hydrogen are expected to play a key supporting role[474,475]. But these technologies are not yet available[476].

Ships and aircraft are significant emitters of greenhouse gases[477]. They must carry heavy loads and travel long distances. Liquid fossil fuels have a major advantage for these applications because of their high energy densities and low cost[478]. The proposed alternatives such as batteries, hydrogen, and biofuels have practical limitations and high costs. These technologies are still in the demonstration or prototype stages.

Several key industrial processes require high temperatures[479]. Examples are cement and metals production. Fossil fuels are used by such processes presently[480]. Alternative low carbon technologies are not yet available on the market.

Historically, new energy technologies have required several decades to move from a demonstration or prototype stages to a wide use[481]. It takes a long while to move from the demonstration stage to when it will be available on the market. It takes even longer for the technology to become competitive once it is available on the market. A technology is available for wide use only when it is competitive to the technology it is replacing. Clearly, many key low carbon technologies will not be available for wide use for a long time[482].

The Transition will Lower Energy Costs

Activists tout that the energy transition will lower the overall costs related to energy. This misinformation is driven by poor estimates that ignore basic science.

These estimates use speculative assumptions. Examples of assumptions are a) the future cost of solar and wind power, b) the cost of the electrical grids which will be of unprecedented size and complexity, c) the level of redundancy required to ensure the stability of electrical grids and d) the future cost of low carbon technologies that are still not available on the market.

Let us examine the problems with such assumptions.

Use of solar and wind power is currently tiny compared to the proposed use in a net zero world[483,484]. According to IEA, electricity from solar and wind power is expected to increase by a factor of fifteen in a net zero world[485]. All grids will have a very high share of solar and wind power. Limited data is available about the issues that would arise from such an extraordinary increase.

This raises several questions. Will the costs of solar and wind power continue to drop and by how much? What will be the impact on the grid stability for each region? What level of grid expansion is practical? How much long duration energy storage will be required?

Several assumptions are required to answer these questions because of the lack of actual data.

The data is also limited for several key technologies. According to IEA, the technologies that will be required to reduce over a third of CO_2 emissions in the net zero world are still not available on the market[486,487]. These technologies address the sectors that are most difficult to decarbonize. Major challenges will need to be overcome before they can be widely used.

There are important unresolved questions. Can these low carbon technologies achieve a cost and performance that is comparable to the conventional technologies? How quickly will the costs drop?

The future costs have many unknowns because of the limited data. From basic math, we know that the use of large extrapolations based on limited data is not reliable. Yet, some studies use numerical values derived from such extrapolations as assumptions in their estimates. Depending on the approach used, a wide range of numerical values can be assigned to the assumptions. Such estimates are highly speculative and are exposed to the bias of the estimator.

The assumptions have a very high uncertainty. Depending on the values used for the assumptions the actual costs could be many-fold higher than the estimated costs or vice versa. The values used for the assumptions are subjective. So, an aggressive value of one estimator can be conservative to another. The potential for bias is very high. Clearly, such cost estimates are not credible.

The problem with assumptions is nicely expressed by this proverb[488]. "*Assumptions are made, and most assumptions are wrong*".

Fortunately, an alternative approach is available to robustly compare the costs between the fossil fuel and low carbon energy system. It does not require assumptions. It only relies on basic science.

The basic science approach was detailed in an earlier chapter. Here, we will consider the highlights.

The cost of a technology to make any product depends on the resource intensity of the technology. The technology with a lower resource intensity has a lower cost to make the product. Examples of resources are labor, materials, land, and energy. Resource intensity informs us about the intensity of the total resource requirements.

It is markedly easier to provide 24X7 electricity from fossil fuel power compared to solar and wind power. This is because nature has provided massive help to fossil fuels. The massive help from nature has ensured that fossil fuel power has a low resource intensity for providing 24X7 electricity. Nature has not provided such help to solar and wind power. Therefore, solar and wind power have markedly higher resource intensities.

The massive help from nature ensures a markedly lower cost of electricity from fossil fuel power compared to solar and wind power. So, the cost of electricity will increase markedly in a net zero world. This has significant implications for electrification.

Electrification is the replacement of technologies that use fossil fuels with technologies that use electricity. Examples are electric cars and heat pumps. Technologies that use electricity have much higher upfront costs. As discussed earlier, the electricity costs will be markedly higher in a net zero world. The higher upfront and higher electricity costs imply that electrification will have higher lifetime costs[489,490,491].

But what about the higher energy efficiency of electrification?

The much higher energy efficiency associated with electrification will markedly decrease energy use compared to fossil fuel technologies. But the lifetime cost will be higher for electrification because of the higher upfront cost and the higher electricity costs. Current electricity prices are low because of the large use of fossil fuel power. Despite the low costs, electrification options have comparable lifetime costs. The much higher electricity costs in the net zero world will ensure a higher lifetime cost for most of the options.

Here is a brief summary. Basic science dictates that the electricity costs will increase markedly with the shift to solar and wind power. This will have a strong negative impact on the lifetime cost of technologies that use electricity. The lifetime cost of electrification will be higher. Overall, the transition will substantially increase the costs related to energy for the global population.

High-Impact Breakthroughs are Common in the Energy Sector

Media regularly touts the breakthroughs reported in the energy field. This gives the impression that breakthroughs will rapidly transform the energy sector. This false belief results from the confusion between a breakthrough and a high-impact breakthrough.

A breakthrough is any major scientific or technical advance. It does not have to satisfy any other criteria. Three million R&D articles are published each year[492]. Thousands of these claim a breakthrough[493].

A high-impact breakthrough is a special type of advance that greatly speeds up the wide acceptance of a new technology. This factor separates a high-impact breakthrough from a breakthrough[494].

Only high-impact breakthroughs are relevant from the viewpoint of transforming the energy sector.

A high-impact breakthrough must satisfy strict criteria. This is essential to speed up the wide acceptance of the technology.

First, there must be a marked improvement in at least one of the following attributes: performance, life, or cost. These are critical attributes for the wide acceptance of an energy technology.

- Performance relates to the ability of the technology to provide the desired function safely and conveniently.
- Life relates to its robustness and sustainability over a long-term[495].
- Cost includes the upfront and total lifetime cost.

The second criterion is far more challenging. The new technology cannot have an inferior performance, life, or cost relative to the current technology. There can be no compromise regarding any of the three critical attributes. If the new technology is inferior in any of these attributes, it does not qualify as a high-impact breakthrough. Such a technology will not be able to organically replace the existing technology on a wide scale[496].

A high-impact breakthrough can be verified only after the technology has demonstrated a major commercial success. The new technology has to be deployed for at least 5 to 10 years to obtain the required data for verification.

High-impact breakthroughs are relatively rare. This is because the technology improvement efforts often have a negative impact on at-least one of the critical attributes. For example, efforts to improve performance are likely to result in an inferior life or cost.

Let us consider the efforts related to solar technology as an example.

Solar cells have been one of the most active areas of R&D. Most efforts have focused on replacing silicon with superior materials. Silicon is the most widely used material in solar cells[497,498]. It has remained so since several decades. The massive R&D efforts have led to hundreds of breakthroughs. These efforts have been widely touted in the media. Several new materials such as perovskites, cadmium telluride, multi-junction, organic photovoltaics have shown promise. But these breakthroughs do not satisfy the criteria for a high-impact breakthrough. The new materials remain inferior in at least one of the critical attributes. For example, perovskites have showed a potential for high performance and low costs but have struggled with the life attribute[499].

History informs us that breakthroughs are common in the energy field. But high-impact breakthroughs are rare. Many breakthroughs have been reported in the solar and bio-energy fields. Most of the solar and bio-energy commercialization efforts based on such breakthroughs have failed[500,501,502]. Why? Because most breakthroughs are not high-impact breakthroughs[503].

A Rapid Energy Transition is Not Too Difficult

Activists believe that a rapid shift to low carbon energy is straightforward. They argue that the only major hurdle is the lack of political will. They do not believe that a rapid transition will be a massive challenge. As discussed below, this belief is false.

The previous transition involved a shift from traditional biomass to fossil fuels. A robust approach to understand the challenges of the low carbon transition is to compare it to the previous energy transition.

An energy transition has three key attributes. They are speed, size, and convenience of energy use. These attributes define the level of the challenge for the transition. Let us look at how the low carbon transition compares with the previous transition for each of these attributes.

Speed is a key attribute because it defines the rate at which new resources will be needed to support the transition. Currently, low carbon technologies contribute to less than 20% of the global energy[504]. Several proposals have a target of reaching net zero emissions by the year 2050[505,506]. A major shift to low carbon energy–from less than 20% to over 80%–is being proposed over the next three decades. For reference, the share of low carbon energy has only increased from 13% to 18% over the last three decades[507,508].

How does the proposed speed compare with the transition from biomass to fossil fuels?

Fossil fuels had a 20% share of the global energy in the mid-nineteenth century,[509]. It took a hundred years to increase the global share of fossil fuels to above 80%. So, the net zero proposals are endorsing a speed that is over three times faster than the previous transition.

The low carbon transition also differs radically in terms of the size or the amount of global energy use. This key attribute defines the quantity of resources required for the transition. The energy used currently is over ten times the energy used at the mid-point of the previous transition[510,511].

We use energy in different forms. This makes it difficult to get a sense of the total amount of our energy use. To get a physical sense it helps to discuss the energy use in terms of tons of oil equivalent. One ton of oil equivalent is equal to the energy available in one ton or 1000 kilograms of crude oil[512]. This is useful because it enables a discussion in units that we can easily grasp. Currently, we use over 10,000 million tons of oil equivalent of energy per year[513,514].

How does this compare with three widely used materials? The combined global use of plastics, steel and cement is 6500 million tons per year[515,516,517,518].

The other way to physically understand the amount of energy use is to compare it with the total food consumed. The global population consumes an enormous amount of food. We can visualize the amount of food we need by the amount of land required to produce the food. 12% of the global land surface is allocated to growing crops and 26% to grazing livestock[519]. So, almost 40% share of the global land is used to meet our food requirements.

How does the global food consumed compare to energy use? The global society consumes over ten times more energy calories than food calories[520].

The low carbon transition differs from the prior energy transition not only in terms of speed and size but also in terms of the convenience offered to society. Convenience is one of the key attributes of an energy transition. A superior convenience leads to a higher acceptance by the consumers. Fossil fuels can provide plentiful energy in a manner that is easy to use. The shift from traditional biomass to fossil fuels markedly increased the convenience of energy use. Such an advantage does not exist for low carbon energy. The best-case scenario for low carbon energy will be to provide comparable convenience to fossil fuel technologies.

A direct comparison with the previous transition reveals the extraordinary challenges for the low carbon transition. The proposed speed for the low carbon transition is three times faster, the size is over ten times larger, and there is no advantage from a convenience viewpoint.

IEA has detailed a pathway to net zero by 2050[521,522]. The report discusses several specific challenges.

- The annual electricity generation will need to increase by 48,000 TWh in less than thirty years. For comparison, it increased by 6000 TWh in the last ten years and by 17,000 TWh over the past thirty years[523,524]. For reference, 1 TWh = 10^9 kWh.
- The global electrical grid networks will need to more than double over the next thirty years. It took over one hundred years to build the existing networks.
- Solar and wind power combined will need to generate 55,000 TWh of electricity annually by the year 2050. Currently, they generate only 3500 TWh of electricity per year[525]. In 2022, the electricity produced from all sources combined was 29,000 TWh[526].
- Battery storage will need to increase to 4200 GW by 2050. It was less than 50 GW in 2022. For reference, 1 GW = 10^6 kW.
- The annual production of low carbon hydrogen will need to increase to over 400 million tons by 2050. The current production of low-carbon hydrogen is negligible[527].
- The CO_2 captured from fossil fuels, industrial processes and bioenergy will need to increase from 45 million tons in 2022 to 5000 million tons by 2050. The CO_2 captured directly from air will need to increase from 0 to 1040 million tons.
- The share of low carbon steel and cement will need to increase from 0% to over 90% by 2050.
- The current share of clean vehicles in the total vehicle sales is low[528]. The share was 13% for cars and 4% for electric buses in 2022. This share will need to increase to 100% by 2050. For freight trucks, the share of total sales will need to increase from 1% to 100%.
- The electric vehicle public charging points will need to increase to 31 million by 2050. There were only 3 million in 2022[529]. Hydrogen refueling stations will need to increase from 1000 in 2022 to 46,000 by 2050.

- The share of low carbon ships will need to increase from 0% to 85% by 2050. The share of low carbon air travel will need to increase from 0% to 70%.
- The share of steel and cement produced by low carbon pathways will need to increase from 0% to over 90% by 2050. The share of chemicals produced by low carbon paths will need to increase from 2% in 2022 to 93% by 2050.
- The share of building that can support zero carbon will need to be higher than 80%[530]. It is currently less than 5%.

Most low carbon technologies are currently deployed at trivial levels. A gigantic and rapid increase is proposed. The proposals require an unprecedented level of resources in a very short period.

Most low carbon technologies use several times more materials compared to fossil fuel technologies[531,532]. The large need for critical minerals is a key issue. Offshore wind power requires fourteen times more critical minerals compared to fossil fuel power per unit of electricity produced over the lifetime of the power plant[533,534]. Onshore wind power requires twelve times more and solar power requires sixteen times more.

An electric car requires six times more critical minerals than a conventional car. A heat pump requires seven times more critical minerals than a gas boiler.

Critical minerals are required for important applications and are at risk for supply disruption[535]. The low carbon technologies require a range of critical minerals. Batteries use lithium, cobalt, and nickel. Wind turbines and electric engines use rare earths (e.g., neodymium). Solar cells require high purity silicon. Electrolysis and fuel cell technologies use platinum or nickel.

Several reports are available on this topic[536,537,538]. Let us look at a few highlights from the recent IEA reports related to the net zero by 2050 scenario[539,540].

- Low carbon technologies will require three times more copper, four times more cobalt, eight times more neodymium, eleven times more nickel, and fourteen times more lithium in the year 2030 compared to what was required in the year 2021.
- An additional production of over 10,000 kt (kilotons) of copper will be required in 2030 to satisfy the increase in demand from the low carbon technologies[541].

- For some critical minerals, the demand from the low carbon technologies alone in 2030 will exceed their total global production in 2021. Such an increase in demand is extraordinary.
- The total global production of lithium was 110 kt in 2021[542]. The demand for lithium will be over 650 kt from the low carbon technologies alone by 2030.
- The total global production of nickel was 2700 kt in 2021[543]. The demand for nickel will be over 3100 kt from the low carbon technologies alone by 2030.
- The total global production of cobalt was 165 kt in 2021[544]. The demand for cobalt will be 200 kt from the low carbon technologies alone by 2030.
- The demand for neodymium in 2030 from low carbon technologies alone will almost be equivalent to its total global production in 2021.

A phenomenal increase in the access to critical minerals, land and skilled workforce will be required at an ultra-fast pace[545].

The proposed transition does not improve the convenience of energy use. Advances in low carbon technologies are necessary to maintain a similar level of convenience. The previous transition was fueled by an enormous improvement in cost and convenience of energy use. Moreover, it was ten times smaller. And yet it took a hundred years! This is a massive challenge for the proposed transition.

An energy transition requires extremely large investments. Such investments are not easily accessible over a short period because of the competition from the many other crucial needs of society. That is one reason why the previous transition took around a hundred years despite the many advantages offered by fossil fuels over traditional biomass. So, access to the enormously large investments in a compressed time frame is another major challenge for the proposed transition.

The technologies that will be required to reduce 35% of the CO_2 are still in the demonstration or prototype stages[546,547]. Moreover, these technologies address the sectors that are most difficult to decarbonize. They represent the most complex and challenging technologies. That is why they are the last to be commercialized.

This further illustrates the extraordinary nature of the challenge for a rapid overhaul of our energy system.

There is a tendency to use our prior successes to downplay the challenge. One example is the touting of the success of our missions to the moon. But this is a very poor analogy. The scope and complexity of sending a few astronauts to moon is trivial compared to the proposed transition.

What would be a better analogy in terms of scope and complexity? The ability to send millions of humans to the moon every year.

There is significant confusion about our past achievements. Our society has made major advances in information technology, medicine, electronics, satellites, and space travel over the past few decades. But these advances required trivial resources compared to the proposed low carbon transition. Recall, we use more energy than plastics, steel and cement combined. The proposed transition involves a complete overhaul of the global energy system. The scope and complexity of the proposed transition is extraordinary because it very rapidly requires an unprecedented level of resources. Global society has never undertaken a task of such scope and complexity. Not even remotely close!

There Will Be No Severe Environmental Issues

The view that the low carbon transition will not cause severe environmental issues is popular[548]. It is mainly based on the limited data that is available for the low carbon technologies. It ignores basic science and the lessons from historical data. This has led to confusion about the future impact from low carbon technologies.

We use technologies such as solar and wind power and electric vehicles at very low levels currently. Solar and wind power provide only 5% of the global energy. Electric cars have a less than 3% share of the total global cars. Experience has taught us that the true environmental impact of a technology is not evident at low levels of use. The true impact is only clear after the technology use reaches a threshold level. This threshold level of use is required to obtain concrete evidence about the true impact.

Let us look at a few examples.

Chlorofluorocarbon or CFC technology was developed in the 1920s for the refrigerant industry. This technology solved the serious health and safety issues related to the prior technology[549]. Media hailed it as a miracle technology[550]. The CFC technology was used across the globe for a long time before its true environmental impact was identified. The true impact–loss of ozone in the stratosphere and the related issues–was only understood after several decades[551]. It took five decades for its use to reach the threshold level required to obtain concrete evidence about its impact[552]. A global agreement was finalized in the year 1987 to phase out CFC technology[553].

The modern plastics technology was invented in the early 1900s. The annual use of plastics increased from two million tons in the 1950s to over four hundred million tons in recent years[554]. The severe impact of plastics on the environment only became clear in the recent decades following a large rise in its use. Most plastics are disposed in landfills, dumps or in the environment after their use[555]. The massive disposal of used plastics has caused severe impacts. Examples of the type of impacts are[556]: a) blockage of waterways and harmful impact to the ocean ecosystem, b) clogging of sewers which leads to an increase in the pests that cause diseases, c) blocking of airways and stomachs of hundreds of species, d) transfer of toxic materials from plastics to humans via the food chain, and e) release of harmful gases during disposal of plastic waste by burning in open air pits. Many countries are taking steps to mitigate the impacts[557]. Again, the harmful side-effects of plastics were not recognized until a threshold level of use was reached.

In the early 2000s, the European Parliament encouraged the use of vegetable oils such as palm oil to decrease the use of fossil fuels[558]. This prompted a large demand for palm oil. Producing countries such as Indonesia increased their palm oil crop in response. The impact of this action became clear only after a decade of accelerated growth of palm oil crop. It led to deforestation, biodiversity loss, and a net increase in greenhouse gas emissions[559]. The European Parliament took corrective actions based on these findings in 2017[560].

The most severe impact from fossil fuels is climate change. But this impact was only recognized after decades of massive use of fossil fuels. Use of fossil fuels began in the 1750s. It took a very long time to establish the link between fossil fuels and severe climate impact. Arrhenius in the early 1900s and Callendar a few decades later were the first to link fossil fuels to the warming of earth. But both believed that the warming would be beneficial in the long term[561,562]. The scientific society did not have any concern about the climate impact of fossil fuels until the middle of the nineteenth century[563,564]. A significant concern emerged only in the early 1970s.

Why was the potential for a serious climate impact not recognized earlier? Because the needed data was not available at low levels of fossil fuels use. Related details were provided in a previous chapter.

Historical data shows that the true environmental impact can only be understood when the technology's use reaches a threshold level. This has very important implications for the low carbon transition. Solar and wind power are crucial components of the low carbon transition. The current level of solar and wind power use is very low. It is several times lower than the level of use at which a significant concern emerged for fossil fuels[565].

How low is the current use of solar and wind power relative to the historical use of fossil fuels?

It is even lower than the use of fossil fuels when their climate impact was believed to be beneficial[566,567,568]. Clearly, the current use of solar and wind power is far below the threshold level at which their true impact can be recognized.

Currently, solar and wind power provide 3500 TWh of electricity. This needs to increase by a factor of fifteen or more according to the IEA and other net zero proposals[569,570].

Will solar and wind power have a severe environmental impact when there is a gigantic increase in their use? Basic science tells us that the answer is a very likely yes.

I have discussed the basic science earlier. I will summarize here.

Fossil fuel power has a major advantage over solar and wind power. Nature has eliminated the diluteness and intermittency challenge of the primary source in case of fossil fuels. Producing 24X7 electricity is much easier for fossil fuel power because of the massive help provided by nature.

Solar and wind power do not receive such help from nature. They must deal with both the diluteness and intermittency challenge. This results in a higher resource intensity for solar and wind power[571].

Because of the lack of help from nature, the resource intensity is much higher for solar and wind power compared to fossil fuel power. Resource intensity tells us about the intensity of the total resources that will be required. Examples of resources are critical minerals, steel, cement, other materials, land, energy, and water. The impact on the environment depends on the resource intensity of the technology[572,573,574]. A high resource intensity equals a high risk for environmental impact.

The resource intensity of solar and wind power is high. This indicates a high possibility for a severe environmental impact[575].

The shift to low carbon power will require an extraordinary increase in solar and wind power and electrification. It will also require a gigantic increase in energy storage technologies and vast electrical networks.

Solar and wind power have a much higher resource intensity than fossil fuels to produce 24X7 electricity. So, low carbon power when used at levels comparable to fossil fuel power will have markedly higher intensity of resource requirements.

This is true for most low carbon energy options. Low carbon energy options require several times more critical minerals and bulk materials compared to conventional options **(Figure 12.1)**[576,577]. Overall, a shift to low carbon energy will require a gigantic increase in the use of critical minerals, land, cement, steel, and other materials[578,579,580]. This suggests a high potential for a severe environmental impact.

Resource Requirements: Critical Minerals & Bulk Materials

Power Sector: Comparison with Fossil Fuel Power Per Unit of Electricity Produced

- Solar power: **16 times** more critical minerals & **8 times** more bulk materials
- Onshore wind power: **12 times** more critical minerals & **5 times** more bulk materials
- Offshore wind power: **14 times** more critical minerals & **4 times** more bulk materials

Electrification: Comparison with Conventional Options

- Electric cars: **6 times** more critical minerals than conventional cars
- Heat pumps: **7 times** more critical minerals than gas boilers

Figure 12.1 Comparison between fossil fuel and low carbon alternatives in terms of their requirements for critical minerals and bulk materials. Bulk materials include steel, aluminum, cement, and plastics. The data for power plants is provided in terms of materials required per unit of electricity produced over the life of the power plant. Data Source[581,582]: IEA

Critical minerals can be used as an example to discuss environmental impacts. Critical minerals are much more difficult to extract, and process compared to common metals. A few issues related to nickel, cobalt, and lithium are discussed below[583].

First, critical minerals have a very high rock-to-metal ratio compared to common metals. This ratio gives information about the quantity of ore mined and waste rock removed to produce a unit of refined metal. It is a critical metric to understand the waste from mines and environmental burdens[584]. Compared to iron and aluminum, the rock-to-metal ratio is 31 times higher for nickel, 108 times higher for cobalt and 200 times higher for lithium[585].

Second, critical minerals require markedly more water. The water use is 106 times higher for nickel, 114 times higher for cobalt and 660 times higher for lithium compared to iron and aluminum[586]. Rare earths require even more water.

A high level of difficulty for extraction and processing means a high impact on the environment. The known impacts related to the extraction and processing of critical minerals are soil erosion, soil contamination, biodiversity loss, contamination of water bodies by chemicals, reduced surface water storage capacity, hazardous waste, and air pollution from fine particles[587,588,589,590].

Practical issues are likely to make it worse. A rapid shift to net zero will create an intense pressure to speed up the production of critical minerals. Health, safety, and the environment are likely to be compromised to decrease the cost and increase the speed of the output at many locations. This will markedly increase the risk of severe environmental impacts.

We will end this section with a brief review. Nature has provided a massive helping hand to fossil fuels. The popular low carbon energy options do not have this benefit. This means that the resource intensity for these low carbon options is markedly higher than that for fossil fuels. A higher resource intensity indicates a higher level of resource requirements per unit of energy produced. When low carbon options are used in comparable amounts as fossil fuels, they will require extraordinary amounts of resources. This is an indicator of a high risk for severe environmental impact.

The Transition Will Improve Energy Security

Energy security is crucial to every country. Activists claim that a shift from fossil fuels to low carbon energy will improve global energy security. This is not consistent with facts.

A country is energy secure if it has reliable access to affordable energy. For energy security, it is important that a few countries do not control global access to energy. Political or civil turmoil, natural disasters or poor policies in these countries could lead to major disruptions in energy supply. One or two countries should not be able to severely impact the global energy supply.

Currently, the global energy supply depends on access to fossil fuels. As the energy transition progresses the global energy supply will increasingly depend on the access to the materials required for low carbon technologies.

How do fossil fuels compare with the low carbon technologies?

Global data from the IEA shows that the production of oil and gas has more geographical diversity. The shares of the regions producing oil, gas and coal are shown in **Figure 12.2 (a, b, c)**[591,592,593]. Oil and gas production is spread across several countries. The production of coal is less diverse. China has a 50% share of the global production of coal. But it also uses over 50% of the global coal[594]. Use of coal is being phased out in many countries. So, the large coal share of China is not a concern for global energy security.

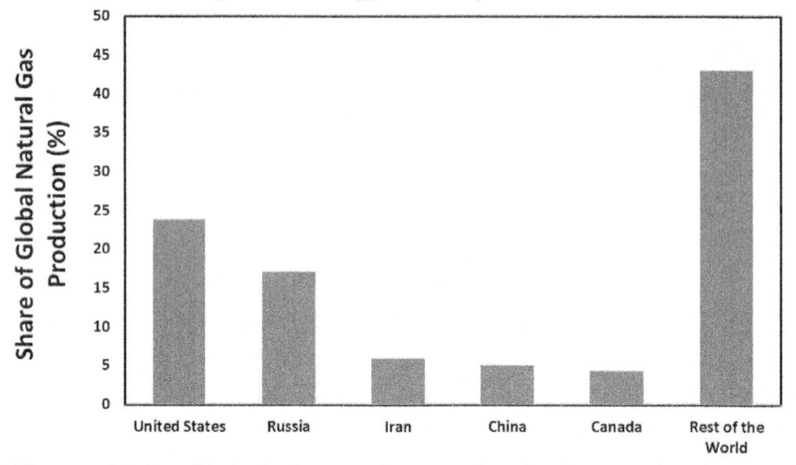

Figure 12.2a Global share of countries in the production of oil. Data is from the year 2021. Data Source[595]: U.S. EIA

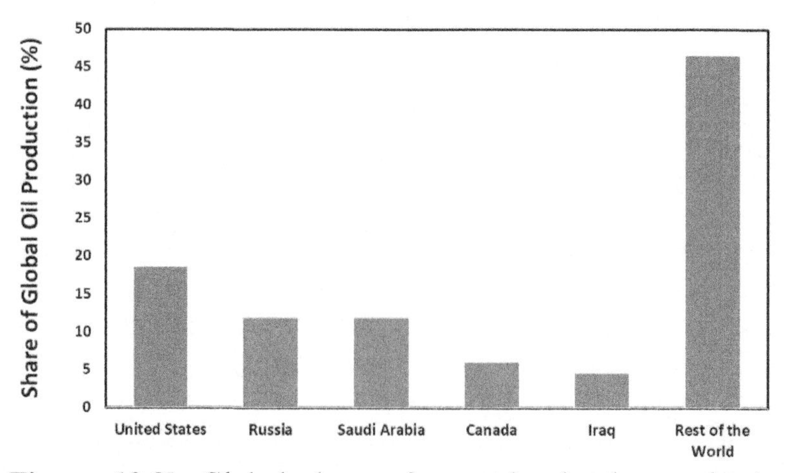

Figure 12.2b Global share of countries in the production of natural gas. Data is from the year 2021. Data Source[596]: U.S. EIA

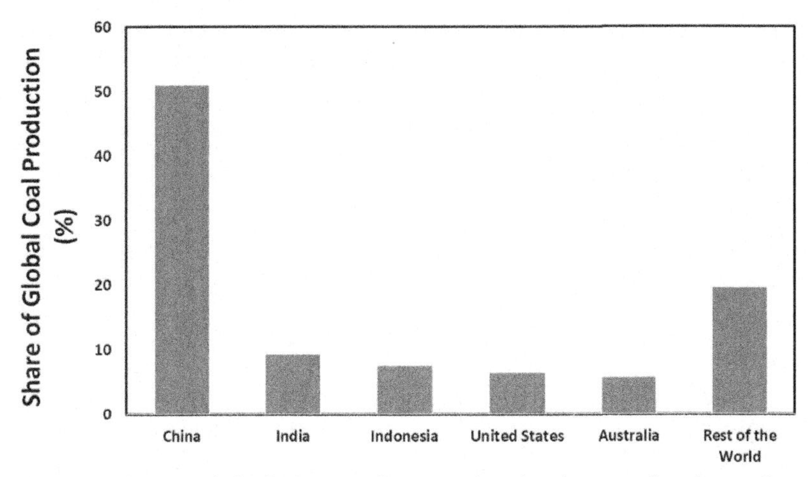

Figure 12.2c Global share of countries in the production of coal. Data is from the year 2021. Data Source[597]: U.S. EIA

Energy security is more challenging for low carbon technologies because they require large amounts of critical minerals, which are geographically less diverse. The current situation about critical minerals and low carbon technologies is discussed in recent IEA reports[598,599]. We will consider the relevant highlights in the next few paragraphs.

The data is startling for most critical minerals. The top producing country controls half or more of the global output for rare earths, graphite, cobalt, lithium, and platinum. China has two-thirds of the share of the global production of rare earths and graphite. The Democratic Republic of Congo has over two-thirds of the share of cobalt production. Australia has half of the share of the global production of lithium. South Africa has over two-thirds of the share of the global output of platinum. Unlike oil and gas, just one or two countries dominate the supply of critical minerals. This is a major concern for global energy security.

The critical minerals require a processing step following their extraction. This step is required to achieve the purity level that is required for final use. The processing of critical minerals is even less diverse. China dominates this space. China controls 60 to 70% of the global processing of lithium and cobalt and 90% of the processing of rare earths. It also controls close to 100% processing of graphite and 80% of the processing to high purity polysilicon.

The overall dominance of China in this space is alarming. It has a large share of the global extraction and processing of many critical minerals. Chinese companies have also made major investments in countries–such as Australia, Chile, Congo, and Indonesia–that have large capabilities for producing critical minerals.

China also dominates the mass manufacturing of low carbon technologies. It has a more than a 70% share in the manufacture of silica-based solar PV modules. Chinese companies have a 55% share in the manufacture of wind turbines and a 65% share of battery cell production.

Such domination of the entire low carbon energy supply chain by a single country is a major concern for the energy security of the other nations.

China has dominated the low carbon energy space because it is uniquely placed. It has large reserves of several key minerals. It has access to a very large, low-cost, and skilled workforce. It was an early mover. Also, it has a political system that can greatly influence investments and the speed of projects. The combination of these attributes is unique. Other nations cannot satisfy these conditions. This tells us that the other nations cannot mimic China's success.

Massive amounts of low carbon technologies will need to be deployed every year for the next several decades during the shift to low carbon energy. Very large deployments will be required routinely even after the shift is complete. Why? Because new deployments will be required to replace the low carbon technologies at the end of their life.

The energy supply of the global society will depend on the supply of critical minerals. Adequate access to critical minerals will be crucial to global energy security.

Can recycling be used to solve the energy security problem? No, it cannot!

An extraordinary increase in the capacity of solar and wind power, low carbon hydrogen and batteries has been proposed[600]. Recall, electricity generation from solar and wind power is proposed to increase by a factor of fifteen. Such an increase will require a gigantic increase in the access to critical minerals.

Solar and wind power plants operate for about 25 years. Thus, recycling will not play a major role in the next few decades. The massive access to critical minerals must be from mining operations in the short to mid-term.

What about the long term? A large recycling of critical minerals is far from guaranteed in the long-term.

The recycling of some critical minerals might not be practical. Certain requirements must be satisfied for successful recycling. A cost and energy efficient process should be available to separate the target material from the other components. A high percentage of the target material should be recoverable.

These requirements are satisfied for certain materials like steel and aluminum[601,602]. But they are not satisfied for many. So, recycling rates are low for many materials[603]. Let us look at two examples.

Recycling of e-waste has been low even though it contains expensive materials. The world produces over 50 million tons of e-waste annually. Only 20% of the e-waste is formally recycled[604].

Accumulation of plastic waste is a severe problem across the globe. Recycling has been considered as a solution for decades now. Many companies continue to make promising claims about the recycling of plastics. After decades of such claims the recycling rates of plastics is still very low. Globally, less than 10% of the plastics is recycled[605]. The recycling rate for plastics is between 5% to 9% in the United States[606,607].

It is very difficult to satisfy the recycling requirements for many materials. This is because of the logistical, technical, and cost challenges. The properties of the material and its application define the level of challenge.

It is not realistic to assume that all the critical minerals will achieve a high recycling rate. Current recycle rates are very low for the materials in solar panels and wind turbines[608,609,610]. Also, less than 2% of lithium is recycled currently[611].

Several companies are making promising claims. Activists get very excited about such claims. But past data warns us to not rely on such claims. The entire process–from collection to reuse–needs to be cost efficient. This must be proven on a large scale over many years. We are at least a decade away from such proof for many critical minerals.

There is a Consensus Amongst Energy Experts

Activists have a wrong notion about who qualifies as an energy expert. This has led to this false belief.

An expert must have the skills and experience to tackle the crucial issues in the chosen field. For example, a climate expert must have expertise in climate science. Expertise is essential to estimate the impacts of climate change. Specifically, a climate expert must have knowledge about climate monitoring tools and the various processes that can influence the climate. An expert must also understand the complex interactions within the climate system. Researchers from universities and other academic institutions have the skills, experience, and motivation to develop such knowledge. So, most climate experts are from the academic world.

Activists recognize the type of knowledge that is required to be a climate expert. But they are not aware of the type of knowledge required to be an energy expert. This is an important gap.

Two questions are crucial in a path forward discussion about the shift to low carbon energy. Which technologies should be used and at what levels? What is the fastest realistic speed for the transition?

To robustly respond to these questions, an energy expert must have knowledge of three key aspects.

- An energy expert must have excellent knowledge about the cost, performance, and limitations of the different technologies. This is crucial to select the optimal technologies and their level of deployment.
- An energy expert must be able to robustly assess the hurdles involved in moving a technology from a development phase to wide deployment. This is important because a large fraction of low carbon technologies is not yet available for wide use. Having a technology on the market does not ensure its wide use. For this, it must show a comparable cost and convenience relative to the technology it is replacing. The technologies required to reduce 35% of the CO_2 in a net zero world are not yet available on the market[612]. It will be a long while before these technologies become competitive for wide use. Moreover, a substantial fraction of the low carbon technologies already available on the market are also not yet competitive. They do not have a comparable cost or convenience to the incumbents[613]. So, these

technologies are also not yet available for wide use. Overall, a large fraction of low carbon technologies is not yet available for wide use. The ability to assess the hurdles for these technologies is crucial to avoid poor assumptions about how quickly they will become available for wide use.

- An energy expert must be well informed about the practical issues associated with massive projects. Examples of practical issues are availability of workforce and other resources, cost escalations, tight schedules, interdependency with other projects, local opposition to the project, getting permits and regulations[614]. These issues are common to all major energy projects. Knowledge of these issues is crucial to understand the practical constraints that can limit the speed of the transition.

Who has the knowledge to qualify as energy experts? Those who have relevant experience with energy technologies. This includes those who have experience with evaluating, developing, and deploying technologies and major projects[615]. This specific experience defines an energy expert. Most of this experience can only be gained in the energy industry. The energy industry has spent hundreds of billions of dollars to develop energy technologies and deploy them over the decades. The massive scale and expansive scope are unique to the energy industry. The energy industry, thus, provides unique access to the relevant experience. So, most energy experts have either a current or past affiliation with the energy industry. It takes a minimum of ten to fifteen years of relevant work in the energy industry to gain such experience[616].

Experience outside the energy industry is less relevant. Universities and other institutions have different goals compared to the industry. Their focus is on advancing science, engineering, and education. Their R&D work is on a small scale, which is very valuable for introducing new ideas. But it is less useful for understanding issues related to large scale deployment. Their overall research scope is narrow and does not address many critical aspects. There is very little opportunity to deploy technologies or undertake major energy projects in an academic setting. So, working outside the energy industry does not offer the same experience to become an energy expert.

This famous quote nicely describes the importance of experience[617]. "*Information is not knowledge. The only source of knowledge is experience. You need experience to gain wisdom.*"

What if many of the contributors to the path forward discussions are not energy experts? This will lead to the use of assumptions that are not realistic. Path forwards based on poor assumptions will cause disruptions in energy supply. It will lead to chaos and undermine the ability of global society to mitigate climate change. To avoid this scenario, a large majority of the people involved with the discussions must have adequate expertise.

But this crucial requirement has been ignored. Researchers who lack energy expertise have dominated the path forward discussions[618]. For example, most of the authors of the IPCC reports on this topic are not energy experts[619]. They do not have the required expertise. Their knowledge about energy is mainly academic because they lack adequate experience in the energy industry.

A large majority of the people who are currently engaged in energy discussions are not energy experts. They do not have practical experience. And yet they are planning the future of our energy[620].

The representation of energy experts is far from adequate in these crucial discussions[621]. This is a recipe for a bad outcome.

Renewables Receive Fewer Subsidies

Activists claim renewables receive fewer subsidies than fossil fuels. As discussed below, this claim has no merit.

Global subsidies data for the recent years is available for fossil fuels. But it is not available for renewable energy.

Fortunately, the relevant data is available for the European Union (EU), and United States (U.S.). This data is useful to compare the level of subsidies.

A recent European Commission report provides information about the subsidies in the EU countries from the year 2015 to 2021[622]. The subsidies changed little on a year-to-year basis. Let us consider the average data. Subsidies for renewables were $84 billion per year. Subsidies for fossil fuels were $59 billion per year. The EU countries used far more fossil fuel energy than renewable energy during the period[623]. The share of fossil fuels was four times more than renewables. Thus, the subsidies per unit of energy consumed in EU countries were several times higher for renewables.

A recent report by the U.S. EIA provides data about the U.S. subsidies from the year 2016 to 2022[624]. The sum of the energy subsidies for all the years during this period was $183 billion. Renewables received 46% of the subsidies. Only 13% of the subsidies were for fossil fuels. 35% of the subsidies were for end use. The major components in the end use category were energy assistance for low-income homes and credits for residential energy efficiency. U.S. consumed eight times more fossil fuel energy than renewable energy[625]. Thus, the subsidies per unit energy consumed in the U.S. were massively higher for renewables.

The Forced Use of Fossil Fuels is Slowing the Transition

Activists claim that the global society is being forced to use fossil fuels, and that is slowing down the transition. I will share key facts that reveal the folly of this claim.

The massive amount of fossil fuel use is being pushed by the needs of our society. Global society has a high demand for fossil fuels because of its unique advantages.

Fossil fuels are a cheap and convenient source of energy. Nature has provided a massive helping hand to fossil fuels[626]. Therefore, fossil fuels have a markedly lower resource intensity compared to renewable energy. Fossil fuels are low cost, easy to produce, store and transport and can provide energy when needed.

Fossil fuels play an important role in the production of plastics, steel, cement, and ammonia[627]. These materials are the building blocks of modern society. Fossil fuels play a major role in every aspect of our life–powering devices and equipment, food, shelter, medical and everyday products. Over six thousand products that we use routinely are made from fossil fuels[628]. Consequently, countries with large access to fossil fuels have prospered relative to those without[629].

The alternatives to fossil fuels are challenged in terms of cost and convenience. Fossil fuels are unique because of the massive support from nature. The basic laws of nature cannot be denied.

We use fossil fuels all over the globe in massive amounts because of their unique advantages and the current lack of suitable alternatives. The global share of fossil fuels in the energy mix has remained above 80% over the past several decades[630]. This is not because of a conspiracy by certain entities to suppress the use of renewables. Nobody can force the use of fossil fuels in our diverse global economy over such long time periods. The needs of our society decide if and how much any product will be used.

It is very important to shift to a low carbon energy system. But this will not happen as fast as desired. Why? Because the low carbon technologies do not have the same level of help from nature. It will be exceedingly difficult for low carbon technologies to quickly achieve a cost and convenience that is acceptable to society. A rapid transition has several extraordinary challenges. These were detailed in an earlier section.

Fossil fuel technologies are still being used extensively because there are no acceptable alternatives for most energy applications. Many key low carbon technologies are still not available on the market[631,632]. The technologies that are commercially available also have challenges[633]. Several decades will be required for the alternatives to become competitive[634], and thereby, acceptable for wide use. Forcing a wide use of the low carbon technologies before they are competitive will lead to a rise in energy poverty.

Our use of fossil fuels is not slowing down the transition. The massive challenges associated with the energy transition are mainly slowing it down. A major reason for the confusion is the misinformation about the energy transition. The challenges are greatly underestimated because of poor assumptions. In the previous section, we saw how the lack of involvement of energy experts is causing a severe problem.

Chapter Highlights

Activists routinely spread false and misleading information about the energy transition. Popular examples of misinformation are listed below:

- Nearly all the technologies required for the net zero transition are already available for wide use.
- The shift from fossil fuels to low carbon energy will lower the overall costs related to energy.
- High-impact breakthroughs are common in the energy sector.
- The proposed rapid shift to low carbon energy is not too difficult.
- A low carbon energy world will not have severe environmental issues.
- The energy transition will improve global energy security.
- Most energy experts agree that the path proposed for the energy transition is robust.
- Renewable energy receives lower subsidies than fossil fuels.
- Global society is being forced to use fossil fuels, which is slowing down the transition.

…§§§-§-§§§…

13. Energy Transition: Misinformation by Skeptics

Skeptics do not see value in the shift to low carbon energy. Their preference is to either stop or delay the energy transition. Here, we will consider the common misbeliefs held by the skeptics on this topic.

A Pure Free-Market System Should Drive the Energy Transition

Skeptics believe governments should not be involved in the energy transition. They propose that a pure free-market system should drive it. I will discuss the problems with this argument below.

A free-market system or economy is voluntary. It is controlled by the laws of supply and demand. The current economic system is largely driven by supply and demand[635].

A free-market system has several advantages. It can efficiently maintain a balance between the supply and demand for goods and services. But it is not equipped to address environmental issues without external intervention. Why? Because a free-market system cannot place a proper price on environmental costs. It undervalues its harmful impact on society.

A free-market system values a product that has a lower cost or a superior convenience. That is not the target when addressing environmental issues. So, a free-market system is not set up to address environmental issues. Some government involvement is necessary.

Let us consider the example of air pollution control in the United States. There was a spurt of economic growth in the United States after the second world war. The increase in industrial activity and energy use caused a major air pollution problem[636]. The air pollution was especially severe in large cities. The market itself did not have the mechanism to address the pollution problem. The U.S. Congress introduced the clean air act in the year 1970[637]. It was designed to protect public health from the air pollution caused by the various sources. The act led to air quality standards that had to be met by each state. The air pollutants targeted were hydrocarbons, particles, nitrogen oxides, sulfur dioxide and carbon monoxide. The clean air act was very successful in lowering the air pollution[638]. It drastically reduced the emissions from power plants, transportation, and industrial activity. For example, new vehicles are 99% cleaner than the 1970 models in terms of the common air pollutants. The clean air act was key to lowering air pollution. Such government involvement is crucial.

Fossil fuel technologies have an advantage because of the massive help from nature. The key low carbon technologies such as solar and wind power do not have such help from nature. I have discussed earlier why the shift to low carbon technologies cannot decrease the cost or increase the convenience of energy use. So, a pure free-market driven system does not have the incentive to drive the shift to low carbon technologies. A limited involvement of governments is necessary[639]. As detailed later, the government policies must be robust. Poor policies will disrupt the energy supply and greatly increase the energy costs.

The Effect of Energy Polices on the Warming is Trivial

Several countries are deploying policies to support the energy transition. Skeptics claim these policies will not have any meaningful impact. They argue that the tiny impact on the global temperature does not warrant the proposed investments. This argument is misleading.

Humans have emitted over 2000 billion tons of CO_2 since the industrial revolution[640]. This has warmed the earth by 1.1°C. This tells us about the relationship between the amount of CO_2 and the temperature rise. A gigantic amount of CO_2 emissions is related to a small change in temperature.

A regional policy that can reduce CO_2 emissions by 0.1 billion ton each year is valuable. Yet, it will have an extremely tiny impact on the global temperature. That is expected based on the amount of the change in CO_2.

The impact of regional policies is small as expected. The cumulative or the sum of the impact of all the regional policies over the next decades is what matters. We can mitigate climate change only if each region plays its part.

It is misleading to focus on an isolated regional policy and claim that its impact on the global temperature is trivial.

Let us consider an example. The U.S. Congress recently approved the Inflation Reduction Act (IRA)[641]. The act includes several policies that support a shift to low carbon energy. The proposed investment is $370 billion. Estimates indicate the policies will reduce the CO_2 emissions in the U.S. by 0.7 billion tons in 2030[642].

Skeptics claim the IRA is a waste of money because it will reduce the global temperature by a trivial amount. But this claim is misleading.

Our energy system is massive. Currently, fossil fuels provide most of the energy[643]. So, a shift to low carbon energy will require large investments across the globe. Such investments are essential to mitigate climate change. Hundreds of trillions of dollars will be needed to target a global temperature that is just a few degrees lower[644,645]. So, an investment of $370 billion will result only in a very tiny change in temperature. There is nothing surprising or alarming about this.

It is misleading to suggest that policies have low worth just because they have a small impact on temperature. A small impact from each region is expected. The sum of the impact from all the regions over the decades is what matters.

Every region will have to take part in setting policies to mitigate climate change[646]. Some regions, such as North America and the European Union, are in a better position to be more aggressive earlier.

In each region, the focus should be on a robust selection of policies. For the IRA or any other proposal, the focus should be on one crucial question. Are the policies consistent with the science and realities of climate and energy?

We should evaluate policies from each region based on the above question. If the policies are consistent with the science and realties of climate and energy, they are robust. Else the policies are poor.

What about the IRA? It is a mixed bag. It has some robust policies. But it also has some policies that are not consistent with the science and realities. There is more discussion on this topic later.

China is the Biggest Contributor to Climate Change

China is the largest emitter of greenhouse gases based on current emissions **(Figure 13.1)**[647]. It produces 60% more emissions than the United States and European Union combined[648]. Skeptics use this as an excuse to delay transition efforts in the western countries.

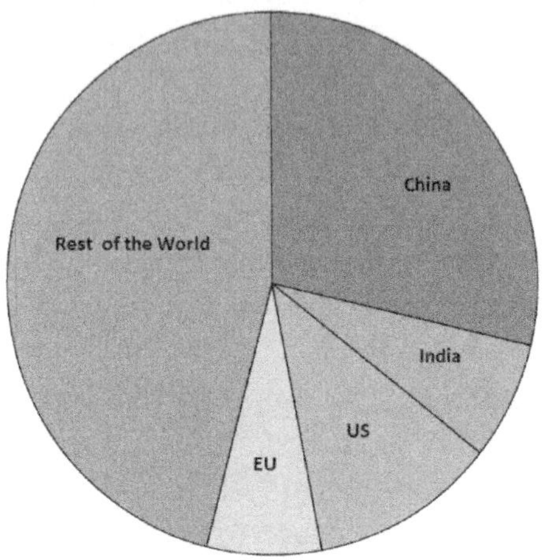

Figure 13.1 Contribution of specific countries to global greenhouse gas emissions. Data is from the year 2021. Data Source[649]: PBL Netherlands Environmental Assessment Agency.

If we were to consider the current emissions alone, China appears to be the biggest contributor to climate change. But this is not the complete picture. While China has been the largest emitter in recent decades, the western countries were larger emitters in the prior decades.

Climate change depends on the content of greenhouse gases in the atmosphere. The greenhouse gases have long lifetimes and accumulate in the atmosphere over time. So, the cumulative contribution is the relevant metric to discuss the relative contribution from countries. Cumulative contribution of a country is the sum of its contribution over all the years since the first use of fossil fuels[650].

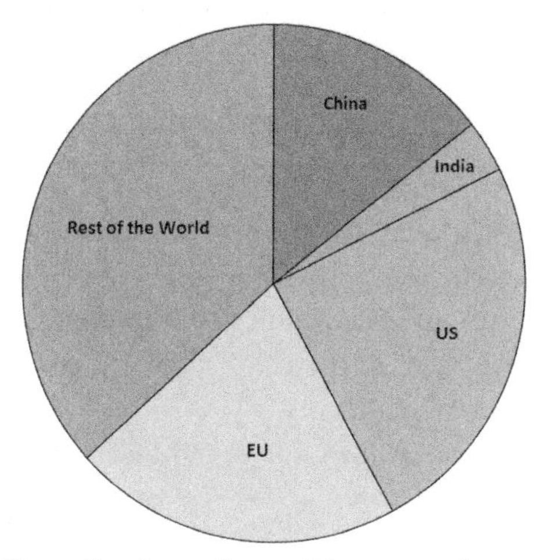

Figure 13.2 Contribution of specific countries to the cumulative CO_2 emissions. CO_2 is used as a proxy for greenhouse gases. Data Source[651]: Global carbon budget.

Figure 13.2 shows the cumulative share of the CO_2 emissions from different countries[652]. The United States is the top contributor. It is closely followed by the European Union. China is a distant third. United States and European Union are larger contributors to climate change than China.

The information about current emissions is also very important. It informs us about the level of reduction that will be required in each country to reach net zero. This will be most challenging for China because of its enormous current emissions.

The cumulative emissions inform us about the total contribution from each country to climate change. United States, European Union and China are the three largest contributors.

The overall data show that the United States, European Union, and China must all show leadership in addressing climate change.

Chapter Highlights

Skeptics routinely provide misleading information about the energy transition. Common examples of misinformation are listed below:

- A free-market system should drive the shift to low carbon energy. Government intervention should be negligible.
- Energy policies are having a trivial impact on limiting the warming of earth.
- China is the largest contributor to climate change.

…§§§-§-§§§…

The Big Picture Discussion

14. Summary of Crucial Facts

The extreme fear of activists and skeptics has created a setting that is controlled by emotions. This emotional setting is a magnet for misinformation. It has created a misinformation crisis.

To move past this crisis, we must be rational and focus on crucial facts about climate and energy. This is essential to develop robust policies, which are key to a brighter future.

What are the crucial facts about climate and energy? They include all the essentials to address climate change in a sustainable way. Specifically, crucial facts include necessary information about the science and realities of climate and energy. Earlier, I used crucial facts to address the misinformation from skeptics and activists. Here, I summarize the crucial facts and briefly discuss the implications of these crucial facts.

Climate Change is a Serious Problem

Our climate system is extremely complex. Tens of thousands of scientists have studied the climate system to better understand it. Climate science has been one of the most intensive areas of research in recent decades[653]. Scientists have had access to an array of weather and climate monitoring tools since several decades[654,655]. These tools have been used to collect a vast amount of data. The data has provided valuable information about the past and the present climate. This information has been used to develop and validate scientific theories. Decades of such efforts have improved the understanding about climate change. Several aspects of climate change are now well understood.

The global temperature has increased by more than 1.1°C since the industrial revolution[656]. The greenhouse gas emissions from human activities have caused this distinct rise in temperature. The atmospheric content of greenhouse gases has increased markedly[657]. The increase has been directly linked to human activities such as the use of fossil fuels[658]. The impact from natural causes has been separated from human causes. Basic physics, the extensive data sets, error analysis, and climate models are all consistent with the conclusion that human activities are the cause[659].

Climate change is having a substantial impact around the globe. Several impacts are understood with high confidence[660,661]. Glaciers, sea ice and ice sheets are decreasing, and sea levels are rising. Plants are growing faster. Oceans are being acidified. Severe events such as hot extremes are increasing. Cold extremes are decreasing. Overall, climate change is having a significant negative impact on the global ecosystems.

The trends in the data, the improved understanding of climate science and the climate models have clarified the future impacts[662]. The impact is expected to become more and more severe as the temperature rises. Climate change will affect a large fraction of the global population over the next decades. It will have a significant impact on life, property, and the ecosystems. The level of impact will depend on the rise in temperature. A higher rise in temperature will also increase the potential for low probability but very high impact events.

The implications are clear. Climate change needs our urgent attention. Strategies to moderate the impacts via climate adaptation alone will not be enough. The more we delay the efforts to reduce our greenhouse gas emissions, the more we will have to adapt. Unwarranted delays will markedly increase the challenges for a good outcome. Both adaptation and a drastic reduction in greenhouse gas emissions will be necessary to lower the short and long-term impact from climate change.

Climate Change is Not the Only Serious Problem

A serious problem is something that deserves urgent attention. Along with climate change, the global population is faced with several other serious problems. For example, poor living conditions are a major cause of severe suffering in the world. Noteworthy progress has been made over the past decades. But the conditions are still dire. Recent data from the United Nations is eye opening.

Many people cannot satisfy even their basic needs.

Two billion people do not have access to safe drinking water[663]. Three billion people have no access to safe sanitation. Billions are expected to lack such basic water services even in the year 2030.

Access to food is also pitiful[664]. Over 700 million people suffer from hunger and over two billion have inadequate access to food.

Access to clean energy continues to be a severe problem[665]. 675 million people do not have access to electricity. Also, billions of people are using highly polluting fuels for cooking and heating. This is causing serious health issues.

Major issues exist with access to healthcare as well[666]. A substantial fraction of the global population does not have access to vital health services. Five million children die before their fifth birthday each year because of poor access to health services, nutrition, and water[667].

It is useful to compare the above data with the recent impact from climate change. Climate change can increase the frequency and intensity of climate disasters such as extreme temperature, storms, floods, wildfires, glacial lake outbursts, droughts, and landslides[668]. The data on such disasters is available from EM-DAT, a widely used database. In recent years, climate related disasters have annually affected 170 million people on an average[669,670,671]. Average deaths have been 0.02 million per year. The current impact of climate change is far less severe than the impact of poverty.

National debts are enormous in most countries[672]. There is massive competition for limited resources. This has critical implications. The efforts to address climate change must not make it more difficult to improve the conditions for the poor. Resources available to address global problems are limited. They must not be wasted via climate policies that are not efficient. Also, there must be no negative impact on the energy access to the poor[673]. If energy access is negatively impacted, it will worsen the suffering of billions of people.

There is No Climate Related Cliff in the Foreseeable Future

There is a lot of panic about natural disasters. The impact from climate disasters has increased over the decades because of climate change. There is a noticeable upward trend over the last several decades. But there is no reason to panic. Many believe that the impact from climate related disasters have been increasing more rapidly in recent years and decades[674]. But that is not true. There has been no unusual recent increase in the impact from the disasters **(Figures 14.1 and 14.2)**[675].

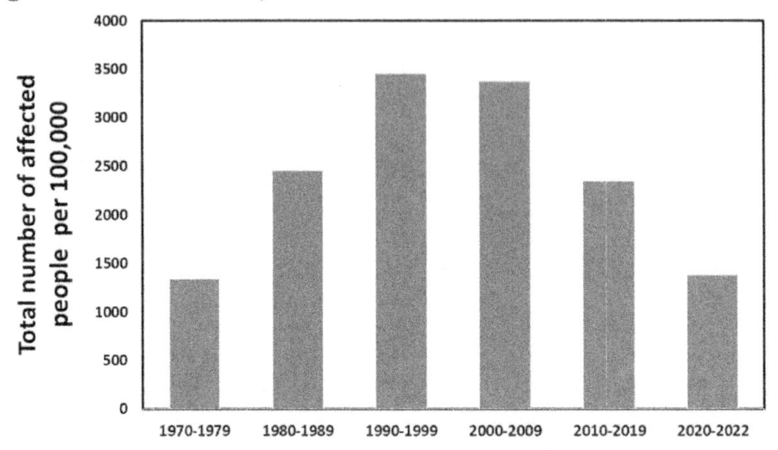

Figure 14.1 Annual average of total number of persons affected by climate related disasters by decade. This data is provided per 100,000 people to account for the change in the global population over time. The total number affected is the sum of injured, those requiring assistance and homeless. Data Source[676]: Our World in Data (EM-DAT, UN world population prospects 2022)

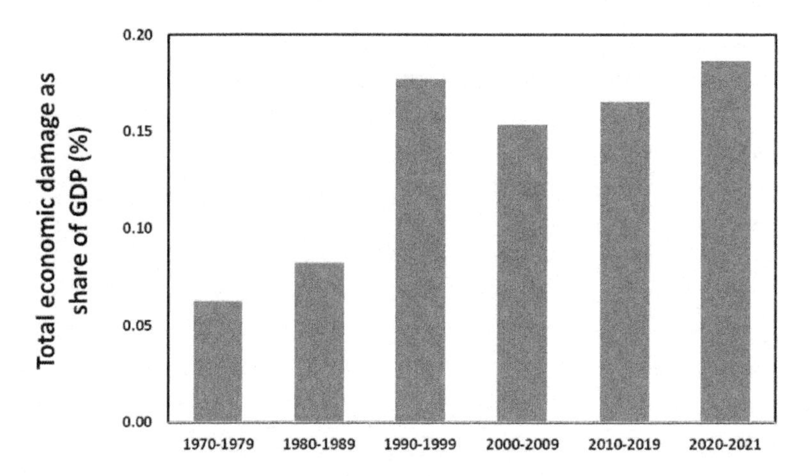

Figure 14.2 Annual average of the economic damage caused by the climate related disasters by decade as a share of GDP. Data Source[677]: Our World in Data (EM-DAT, World Bank)

Most scientists agree that the impact of climate change will markedly worsen with every increment of rise in temperature. The information is summarized in the latest IPCC report[678]. Overall, a far more severe impact from natural disasters is expected for a warming of 4°C by the year 2100 compared to a warming of 1.5°C. But there is no credible evidence to suggest that the natural disasters will cause destruction of the global society for a 4°C warming.

The global data shows that the impact from climate disasters has not risen unusually in the most recent decades[679]. Notably, the current impact from climate related disasters is far less than the impact from poverty[680].

How much of a temperature rise are we headed to by the end of this century?

The temperature is expected to rise between 2.5°C to 2.9°C according to a recent report by the United Nations[681,682]. This estimate is based on the current policies and commitments of the countries. The International Energy Agency (IEA) estimates a temperature rise of 2.4°C in their stated policies scenario, which includes current policies and those under development[683].

Realistically, global governments are expected to improve beyond their current policies in the long-term. For example, according to announced pledges scenario of IEA, which includes short- and long-term national pledges, the global temperature is estimated to rise to 1.7°C[684]. So, it is reasonable to expect that the temperature rise will be markedly lower than 2.5°C by the year 2100. There is no reason to worry about society falling off a cliff because of climate change in the foreseeable future.

There is excessive fear amongst many activists about the occurrence of very high impact events that would destroy human society. The scientific community agrees that the probability of such events is low in the foreseeable future[685,686]. Clearly, excessive fear about such events is not warranted because of the low probability.

Consider the fact that there is a range of outcomes for many of our daily actions. For example, each time we travel there is a small possibility of an accidental death. But we do not focus on this negative possibility. Why? Because the possibility for that outcome is low.

Similarly, a realistic view about the impact of climate change is important[687]. A recalibration to a realistic view can prevent a push for energy policies that are not sustainable. It can also prevent severe mental anguish, especially amongst the younger population.

Economic State of a Country Depends on its Energy Use

The economic state of a country defines the standard of living of its population. The standard of living in a country can be numerically discussed in terms of its gross domestic product (GDP) per capita[688]. GDP is the value of goods and services produced in the country. It is commonly used to discuss the economic state of a country. Developing countries have a much lower GDP per capita compared to developed countries and thereby a markedly lower standard of living.

Most countries have increased their GDP per capita by increasing the output from farming and factories[689]. They have achieved an increase in output by using machines to mechanize the operations. A large amount of energy is required to power the machines.

Therefore, countries have needed a large increase in energy to increase their output.

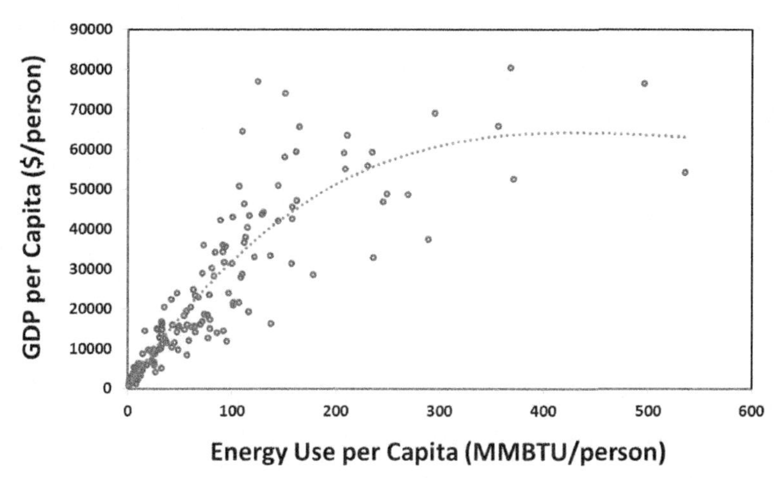

Figure 14.3 Data from more than 170 countries showing the relation between their standard of living (GDP per capita) and energy use per capita. Data is for the year 2021. Data Source[690,691]: The World Bank and U.S. EIA

Figure 14.3 shows the relation between GDP per capita, and energy use per capita of over 170 countries across the globe for the year 2021[692,693]. A strong linear correlation is observed until a certain level of energy use per capita is achieved by a country. I will refer to this as the poverty-buster level. The data shows the poverty-buster level is 80 million BTU of energy use per person. For reference, 1 million BTU = 293 kWh. High-income countries or advanced economies use far more than this level of energy.

The figure shows that the GDP per capita correlates very strongly with energy per capita in the case of low- and mid-income countries.

The human development index also correlates strongly with energy per capita in case of low- and mid-income countries[694,695]. This index includes key dimensions of human development such as a long and healthy life, good education, and a decent standard of living.

India has 1.4 billion people. Its energy use is less than 25 million BTU per person[696]. So, energy use in India is far below the poverty-buster level.

Much of the global population lives in countries where the energy use is below the poverty-buster level of 80 million BTU per person. Even a moderate increase in the energy use in such countries will lead to a substantial increase in their GDP per capita. This means a much better standard of living and human development index.

The case of China and India is useful to discuss the importance of increasing the energy use in developing countries. Both India and China had a similar GDP per capita in 1990[697]. China increased its energy use per capita almost twice as fast as India since 1990. The economic benefit to China was immense. **Figure 14.4** shows the massive benefit to China's GDP per capita because of the faster increase in its energy use.

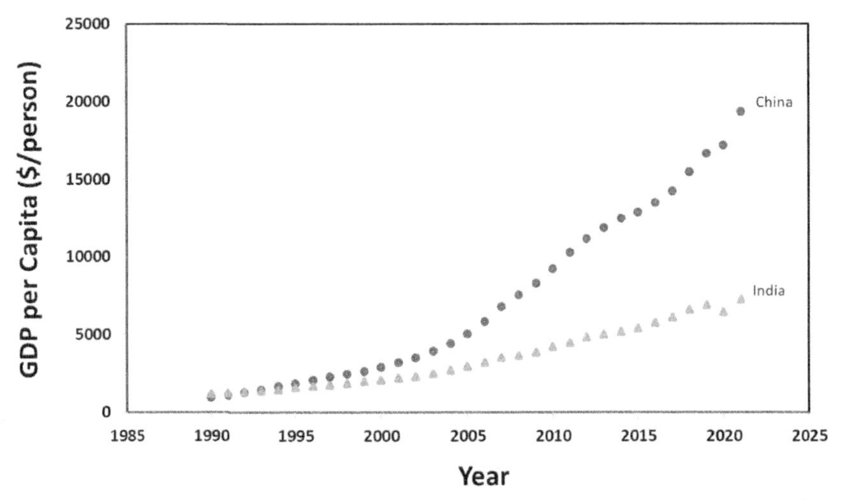

Figure 14.4 Comparison of GDP per capita between India and China since the year 1990. Data Source[698]: The World Bank.

Developed countries have a high GDP per capita because of their high energy use. But many developed countries use far more energy than the poverty-buster level of 80 million BTU per person. The GDP per capita is less sensitive to the energy per capita beyond the poverty-buster level.

The above facts have critical implications. Most developing countries will need to markedly increase their energy use to achieve a decent standard of living for their citizens. More than half of the global population currently lives on less than $10 per day[699]. Most of this population lives in developing countries. Adequate access to low-cost energy is crucial to the progress of developing countries.

The situation is different for developed countries. They already have an excellent standard of living because of high energy use. These countries have an energy per capita that is far more than the poverty-buster level. The standard of living in such countries will not improve much by a further increase in energy use. Countries with a very high energy per capita can decrease their energy use substantially without a major compromise to their standard of living or human development index.

A Perfect Energy Technology Option Does Not Exist

Each energy technology has critical advantages and challenges. There is no perfect option.

Let us first consider the options to reduce greenhouse gases from the power sector.

Cost is a critical attribute for the options in all sectors. Two types of cost are important for this discussion. The upfront cost refers to the initial investment that is required. The lifetime cost refers to the cost over the lifetime of the technology. The cost discussion is based on several credible resources[700,701,702,703,704,705]. For reference, I present the upfront cost and lifetime cost in the United States for the different power options in Figures 14.5 and 14.6[706,707,708,709,710].

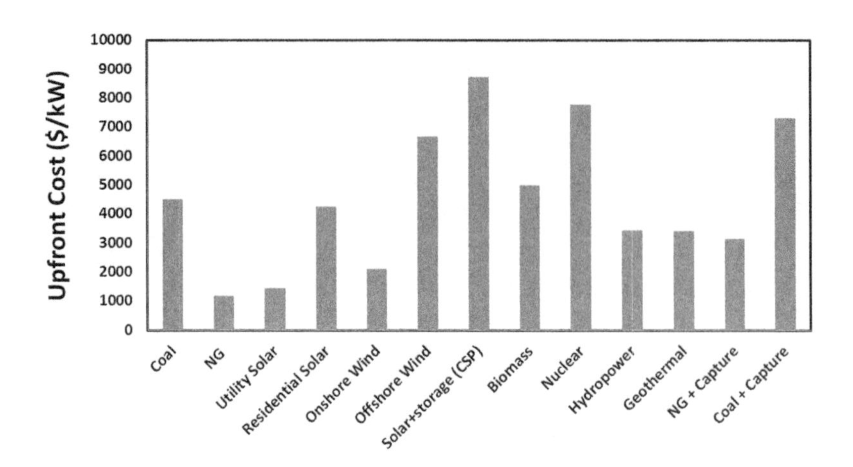

Figure 14.5 Upfront costs for the different power generation technologies in the United States. The costs are in 2022 $/kW. NG = Natural gas. Data Source[711]: U.S. EIA and NREL

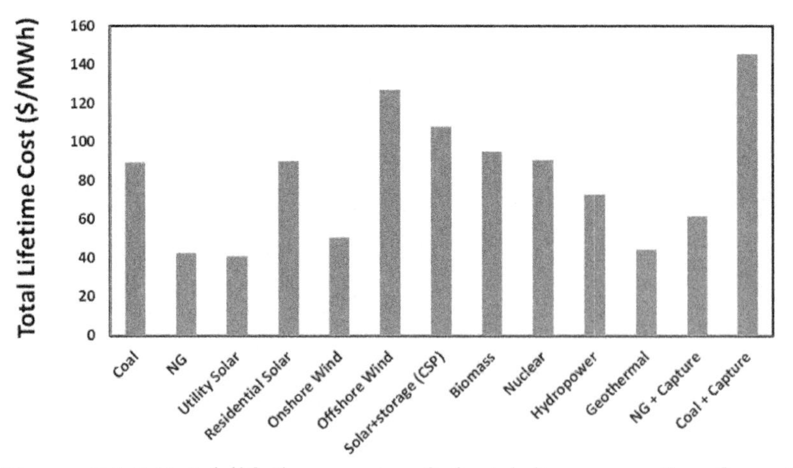

Figure 14.6 Total lifetime costs of electricity generation from the different power generation technologies in the United States. These are average costs and are in 2022 $/MWh. NG = Natural gas. Data Source[712]: U.S. EIA, NREL and NETL.

Utility solar has a low upfront and low lifetime cost compared to other options. But this is only true when it is deployed to a limited extent because of the intermittency challenge of solar power. A utility solar power plant requires more than hundred times the land compared to a natural gas power plant[713]. So, land availability can be a challenge for utility solar near high population centers. Residential solar does not require any extra land. But it is two to three times more expensive than utility solar. Residential solar has a high cost compared to other options even without energy storage. The cost is very high for residential solar with energy storage[714].

Onshore wind also has a low upfront and low lifetime cost. The low costs of onshore wind are only relevant at a limited level of use because it too suffers from an intermittency challenge. Onshore wind plants require several hundred times more land than natural gas power plants[715]. So, availability of land can be a challenge near urban centers. This is not a concern for offshore wind. But the lifetime cost for offshore wind is far higher. In the United States, its lifetime cost is three times that of onshore wind.

The wide deployment of solar and wind power requires long duration energy storage. It can address the drawbacks of solar and wind power. Batteries and hydrogen are popular options for long duration storage. Solar or wind with long duration storage has a high upfront and high lifetime cost. This is because long duration energy storage is expensive.

Hydropower has a midrange upfront and lifetime cost compared to other options. Hydropower is currently the largest source of global low carbon electricity. The flexibility of hydropower makes it an excellent complement to solar and wind power. A suitable location is especially important for hydropower plants. The location of the project determines the cost, environmental and societal impact[716]. There are limited suitable locations across the globe[717]. Many hydropower projects have been completed over the past decades. These projects occupy many suitable locations. So, access to suitable locations for new projects is a significant challenge. Also, hydropower projects have long lead times.

Nuclear power has a midrange-to-high upfront cost and midrange lifetime cost. Unlike solar and wind power, it does not suffer from an intermittency challenge. Nuclear power plants produce radioactive waste. Transportation and safe disposal of such waste requires special handling. Nuclear projects have long lead times.

Biopower has a midrange upfront cost and a midrange-to-high lifetime cost. The lifetime cost is sensitive to the feedstock costs. Like fossil fuel and nuclear power, biopower can also produce on demand electricity. Biomass plantations have large water and land requirements. It is a challenge to transport biomass over long distances because of its low density. So, biopower plants need to be located close to biomass plantations.

Geothermal power has low costs compared to other options when high-quality resources are available. But high-quality resources are scarce. Geothermal power has a high upfront and a high lifetime cost for use with low quality resources[718]. Geothermal power does not suffer from an intermittency challenge[719]. Geothermal power has a much smaller land footprint compared to other renewable technologies.

Natural gas power has a low upfront cost. The lifetime cost is also low for countries with cheap natural gas. But it is midrange-to-high in countries with a high natural gas price. Replacement of coal power with natural gas power can reduce greenhouse gas emissions by 50%. This reduction in greenhouse gases is far inferior to that achieved by renewable and nuclear power. Further reduction is possible by coupling natural gas power with CO_2 capture and storage (CCS) technology. But the CCS technology is not yet cost-effective. Once it becomes so, the natural gas power plants can be retrofitted with CCS to further reduce greenhouse gases. This can reduce greenhouse gases by an additional 70%[720].

Natural gas power plant with CCS has a low-to-midrange upfront cost and a midrange-to-high lifetime cost. The lifetime cost is sensitive to the natural gas price. This technology has a superior cost profile in regions with a low natural gas price. For example, its lifetime cost is midrange in the United States where natural gas is available at a low cost[721].

Coal power has a much higher upfront cost compared to natural gas power. Likewise, coal power with CCS also has a much higher upfront cost than natural gas with CCS. Overall, coal power with CCS has a high upfront cost and midrange-to-high lifetime cost compared to other options. The competitiveness of the lifetime cost depends on the coal price. For example, the United States has a relatively high coal price. So, the technology has a high lifetime cost in the United States. CCS reduces greenhouse gas emissions from coal power plants by 60 to 80%. Coal power plants with CCS have higher emissions than other low carbon power technologies.

Next, let us consider the options to reduce greenhouse gas emissions from the cars sector.

Electric cars have a high upfront cost when compared to the other options[722,723]. The lifetime cost is low-to-midrange[724,725,726,727]. They can decrease greenhouse gas emissions by 50% based on an average global electrical grid[728,729]. The decrease in each region depends on the carbon intensity of its electrical grid. The charging of electric cars is slow compared to refueling of conventional vehicles[730,731]. This translates to a longer wait time. A typical electric car has a much shorter travel range compared to a conventional car[732].

Hybrid cars have a low-to-midrange upfront cost and a low lifetime cost[733,734]. They can reduce greenhouse gas emissions by 20 to 30%[735,736]. Hybrid cars do not suffer from either a slow refueling speed or a short travel range[737].

Advanced biofuels have a high upfront and a high lifetime cost[738,739,740]. They can reduce greenhouse gas emissions by 90%[741]. They do not suffer from either a slow refueling speed or a short travel range.

Fuel cell cars have a high upfront and a high lifetime cost[742,743]. They can reduce greenhouse gas emissions by over 40%[744]. They do not suffer from either a slow refueling speed or a short travel range.[745].

Expansion of mass transit has a low upfront and a low lifetime cost[746]. This option can reduce greenhouse gas emissions by over 60%[747,748]. The level of reduction depends on the vehicle occupancy. This is the only option that can also reduce traffic congestion and accidents. But mass transit is less convenient compared to personal cars. A rapid expansion of mass transit is especially challenging in regions with low population density.

One option which is very controversial is green hydrogen. Currently, it has very high upfront and lifetime costs[749]. The high costs are because of the expensive equipment and a low overall efficiency[750]. Storage and transportation of hydrogen is also a challenge because of its low volumetric energy density and safety concerns. But it also has some interesting properties such as a high energy content per unit weight[751]. It is being considered as a solution for several difficult to decarbonize applications because it is versatile energy carrier[752].

There is no silver bullet option. Extreme passion for any specific option is not justified. The best option will depend on the geography of the region, the existing energy mix, logistics and the time of deployment.

The Energy Transition Will Be Very Costly

I will start this discussion by reviewing the upfront costs of the energy transition. These are mainly the costs of the physical assets that will be required to shift to low carbon energy. The total upfront cost is estimated to be over a hundred trillion dollars[753,754].

The study by the consulting firm McKinsey & Company is useful for this discussion[755]. According to the study, the global society will have to spend an additional $3.5 trillion per year for thirty years[756]. The $3.5 trillion is the spending that will be required over and above the amount that was spent on energy assets in the year 2020.

How much is $3.5 trillion per year? It is equal to half of the annual global corporate profits or one quarter of the annual global tax revenue[757].

The cost estimate in the McKinsey study is based on the physical assets required for the energy transition. It excludes several other costs. Examples of excluded costs are a) retraining of the workforce, b) compensation for stranded assets, c) loss of value pools in parts of the economy, and d) redundancy in energy systems that would be required to avoid supply volatility during the transition.

The excluded costs are expected to be huge considering that a very large workforce will need to be retrained, vast number of assets will be stranded, and substantial redundancy in energy systems will be required[758].

Thus, the cost estimates from the study are conservative. Despite being low-end costs, they are massive.

What about the impact of the transition on the lifetime cost of energy?

The energy transition consists of three major steps: a) a shift to low carbon power, b) electrification and c) decarbonization of the most difficult applications. Each of these steps will increase the cost of energy.

First, I will consider shift to low carbon power.

The current share from solar and wind power is small relative to fossil fuel power[759]. A massive shift to solar and wind power is common to most net zero pathways[760,761]. As discussed earlier, a shift from fossil fuel power to solar and wind power will markedly increase the cost of electricity[762]. A brief review is provided below.

The resource intensity required for making a product defines its cost[763]. A higher resource intensity equals a higher cost. Basic science can inform us about the relative resource intensity required to make electricity from solar, wind and fossil fuel power.

By providing massive help, nature has ensured a low resource intensity for fossil fuel power. Unlike fossil fuel power, solar and wind power receive little-to-no help from nature. So, a markedly higher resource intensity is required to produce 24X7 electricity from solar and wind power. Basic science dictates that the cost of electricity generation will be markedly higher for solar and wind power.

Next, I will discuss the cost issues related to electrification.

Electrification technologies have a high upfront cost. For example, electric vehicles and heat pumps are far more expensive than their conventional counterparts[764]. A low electricity cost is very important to lower the lifetime costs. The cost of electricity needs to remain low to avoid an increase in the lifetime cost of electrification options compared to the conventional technologies. Recall, the shift to low carbon power will markedly increase the electricity cost. The higher electricity cost will markedly increase the lifetime cost of electrification options.

Finally, I will discuss the cost issues of the technologies required to address the most difficult applications.

The technologies required to reduce 35% of the CO_2 in a net zero world are still not available on the market[765,766]. The applications addressed by these technologies are the most difficult to decarbonize. Some examples are ships, air travel, and industrial processes that require a high temperature. Technologies for such applications have far more challenges than the other decarbonization technologies. That is the reason these technologies are not yet available on the market. A high level of challenge equals to a high resource intensity. This equals a high cost.

The upfront energy costs will be gigantic. Also, the lifetime energy costs will increase markedly as the world moves to net zero. The overall cost impact on the global society will depend on the energy policies employed. An optimal selection of technologies at the appropriate time can lower the impact.

Energy Transition Will Have a Severe Environmental Impact

Basic science along with past learnings indicate that a shift to low carbon energy will very likely lead to severe environmental concerns. Let us review the highlights below.

The massive role of solar and wind power is common to most net zero proposals[767,768]. Basic science informs us about the risk of severe impact from solar and wind power. The environmental impact of a technology depends on its resource intensity. The resource intensity of a technology is the overall intensity of all its required resources such as materials, energy, land, and water. There is an impact on the environment each time a resource is required[769,770]. So, a technology with a high resource intensity will have a high impact on the environment.

I have discussed details earlier and will only summarize here.

The resource intensity for solar and wind power is markedly higher than fossil fuel power because they receive far lower support from nature. The high resource intensity of solar and wind power indicates that they have a very high potential for severe environmental impact. Recall, solar and wind power will require massive overbuilding, back-up power, energy storage, and grid extensions to supply 24X7 electricity. So, the impact from all the supporting infrastructure will also contribute to the total environmental impact.

Why is the impact not already recognized? Because we currently use solar and wind power at a very small scale[771]. Their impact on the environment appears to be low because adequate data is not available at a low level of use. The true impact of a technology is only understood after it has reached a threshold level of use. Let us consider the example of fossil fuels.

The severe impact of fossil fuels was only recognized after massive use. The pioneers, who first connected fossil fuels to the warming of earth, believed that the warming would be beneficial in the long term[772,773]. This was the case even in the 1930s when fossil fuels were already used in large amounts. Why was there no concern back then? Because the level of fossil fuels use was below the threshold level to understand the impact. The impact from warming became a major concern only after fossil fuels were used in massive amounts.

Solar and wind power are currently used at lower levels than fossil fuels were used back in the 1930s[774]. Adequate data is not available to understand the environmental impact from solar and wind power because of such low level of use.

The shift to low carbon energy will require an extraordinary increase in solar and wind power. According to the IEA net zero proposal, the electricity from solar and wind power will need to increase by a factor of fifteen[775]. The shift will also require a gigantic increase in energy storage technologies and vast electrical networks. Solar and wind power requires many times more materials and land compared to fossil fuel technologies[776,777,778]. So, a gigantic increase in critical minerals, land, cement, steel, and energy will be required.

The level of the resources used is currently small because of the low levels of solar and wind power use. The true impact will only be understood at much higher levels of use.

Similarly, other low carbon technologies are also being used at a very low level currently[779]. For example, the current share of electric cars is less than 3%[780]. The share of most other low carbon technologies is also very low.

Technologies required to reduce 35% of the CO_2 in a net zero world are yet in the demonstration or prototype phases[781,782]. These technologies have severe challenges. The severe challenges have delayed their deployment. A high level of challenge for a technology is an indicator of a high resource intensity.

Let us end this section with a short summary. Nature has provided a massive helping hand to fossil fuels. The popular low carbon energy options do not have this benefit. The laws of nature dictate that the resource intensity of the low carbon technologies is markedly higher than fossil fuel technologies. Resource intensity is an indicator of the *overall intensity level* of required resources such as critical minerals, other materials, land, fuel, water, and labor. A higher resource intensity for the low carbon technologies means they require a higher intensity level of resources per unit of energy produced compared to fossil fuel technologies. Currently, this is not a problem because low carbon technologies have a small share in the energy mix. But the situation will be totally different in a net zero world. After the shift, low carbon technologies will need to satisfy the global energy needs. At such massive levels of use, the low carbon technologies will require an extraordinary level of resources. This will very likely cause a severe impact on the environment. The level of impact will depend on the path forward. An optimal technology mix and a smart use of energy will be key to lower the impact.

The Energy Transition Has Many Extraordinary Challenges

The proposed shift from fossil fuels to low carbon energy is far more challenging than the prior shift from traditional biomass to fossil fuels.

The amount of energy involved in the low carbon transition is ten times larger than the prior transition[783].

The prior energy transition took place over a hundred years. The low carbon transition is proposed over a period of thirty years. The proposed speed is over three times faster. The ultrafast speed magnifies the challenge from the gigantic size.

Energy drives our world. We use more energy annually than plastics, steel and cement combined. An overhaul of our energy system is an overhaul of how we produce, travel, and live. We can expect such an overhaul at an ultrafast speed to pose extraordinary challenges.

The IEA has provided details about its net zero by 2050 proposal in recent reports[784,785,786]. I will share examples from these reports to highlight the key challenges[787].

The share of low carbon energy is currently less than 20%. It is proposed to grow to 80% in the next three decades. Why is this extraordinary? Because the share of low carbon energy has only increased from 13% to 18% over the last three decades[788,789].

The global electricity networks will need to more than double over the next thirty years. Why is this extraordinary? Because the existing networks took over a hundred years to build[790]. For reference, the total length of transmission circuits and distribution grids is over 80 million kilometers[791,792].

The annual electricity produced from solar and wind power will need to grow from the current 3500 TWh to 55,000 TWh by 2050. Why is this extraordinary? Because the global annual electricity *from all sources combined* only increased by 6000 TWh in the last 10 years and 17,000 TWh over the last thirty years[793].

Battery storage will need to increase to 4200 GW by 2050. This is extraordinary considering that the existing capacity is less than 50 GW.

The current levels of low carbon energy use are very low[794]. An extraordinary growth will be required in all the energy areas by 2050[795].

The annual production of hydrogen from low carbon sources will need to increase from the current trivial amounts to over 400 million tons[796]. The share of low carbon emissions steel and cement will need to increase from 0% to over 90%. The share in the sales of low carbon buses will need to increase from 4% to 100%. The share in the sales of low carbon freight trucks will need to increase from 1% to 100%. The public EV chargers will need to increase from 3 million to over 30 million units. The share of low carbon ships will need to increase from 0% to 85%. The share of low carbon air travel will need to increase from 0% to 70%. The share of buildings that can support zero carbon will need to increase from the current trivial levels to over 80%. The total CO_2 captured will need to increase from 45 million tons to 6000 million tons.

Why is this growth an extraordinary challenge? Because it will require extraordinary amounts of resources at an ultrafast pace. Resources include critical minerals, materials, land, energy, water, and skilled workers.

Let us examine the challenges related to some of these resources.

Low carbon technologies require several times more critical minerals than fossil fuel technologies. Solar power requires sixteen times more critical minerals compared to fossil fuel power per unit of electricity produced over the life of the power plant[797,798]. Onshore wind power requires twelve times more critical minerals while offshore wind requires fourteen times more. An electric car requires six times more critical minerals than a conventional car. A heat pump requires seven times more critical minerals than a conventional gas boiler.

The proposed increase in low carbon energy will lead to a gigantic increase in the demand for critical minerals **(Figure 14.7)**. The situation will be extreme for certain critical minerals such as lithium, nickel, neodymium, and cobalt. By the year 2030, low carbon technologies alone will need these minerals in quantities that are equivalent or more than the current total global production[799,800,801]. Such requirements are unprecedented.

Increase in Critical Minerals Demand for Low Carbon Technologies from 2021 to 2030

- Copper demand will increase by a **factor of three**
- Cobalt demand will increase by a **factor of four**
- Neodymium demand will increase by a **factor of seven**
- Nickel demand will increase by a **factor of ten**
- Lithium demand will increase by a **factor of fourteen**

Figure 14.7 The estimated increase in the demand for critical minerals from 2021 to 2030 based on the IEA net zero by 2050 pathway. Data Source[802]: IEA

Many new mining projects will be required to provide the critical minerals. It takes a long time to produce minerals from a new project. Such projects carry a high financial risk. Recent global data shows an average of twenty years of lead time for a new mining project for some key critical minerals[803]. Exploration takes about twelve years. Construction of the mine and ramping up production requires another eight years.

Rapid access to the massive amounts of the required critical minerals will be a major challenge because of the high financial risk and long lead times for new mining projects.

What about land resources? Solar and wind power plants require far more land than natural gas and coal power plants[804]. A solar power plant requires over a hundred times more land to produce the same amount of electricity as a natural gas plant. A wind farm requires over a thousand times more land. The challenge is not that there is not enough land. The challenge is about the practical access to the required land at the desired locations.

The solar and wind projects often need to be sited close to the population centers for economic and other practical reasons. Many people are opposed to having such projects close to where they live. This is already having a major impact.

Globally, there is an increasing level of such "not in my backyard" attitude. Energy projects are increasingly seeing pushbacks from the local population. This is of major concern to the energy transition because of its extraordinarily large land requirements. New renewables projects and new mines will require vast areas of land.

Let us review the example of United States. Renewable energy projects have faced a substantial opposition in 45 states[805]. Over two hundred local laws and policies have been issued in 35 states to restrict renewable energy projects. Renewable projects are currently being deployed at very slow speeds compared to what is proposed in the future. Even so, the opposition for projects is significant. Several times faster deployment is being proposed for the next three decades. This will markedly increase the severity of this challenge.

The requirement of a very large skilled workforce also presents a massive challenge. Tens of millions of skilled workers will be required for the shift to low carbon energy[806]. A shortage of workers is already being observed at the current levels of deployment. China is struggling to fill positions in its factories. Europe and United States have a shortage of plumbers, pipefitters, electricians, heating technicians and construction workers. Such shortages are already restricting the pace of the shift. This challenge will rapidly increase because of the stress from the ultrafast deployment speeds that are being proposed. China's ministry of education has predicted a huge talent gap by the year 2025. It has predicted a talent gap of over 9 million people in power equipment, over 1 million in clean vehicles and more than a quarter of a million in offshore equipment.

What about technology availability? Technologies required to reduce 35% of the CO_2 in a net zero world are still not available on the market[807,808]. These technologies will be required to address the more difficult applications such as steel, and cement production, ships, aircraft, long duration storage and direct CO_2 capture from air.

These technologies have far more challenges than solar and wind power, heat pumps, and electric cars. That is why they are still in the demonstration or prototype stages. I will call these difficult technologies.

What does history tell us about the time required for a new energy technology to move from a demonstration stage to a relatively small yet significant market share? For wind power it took over 30 years[809,810]. For solar power it took over 50 years. And it took 30 years for lithium battery[811].

Many of the difficult technologies have only been in the prototype or demonstration stages for about a decade. How much time will be required for these technologies to reach a wide scale use? No one can accurately predict this[812]. But we can guess based on historical data. The easier technologies such as solar and wind power have taken between 30 to 60 years to reach just a small fraction of the market share. Therefore, most of the difficult technologies are likely to take several decades more before they have a major share of the market. This is another extraordinary challenge.

Energy security is another major challenge. Geographical diversity of energy resources is crucial for global energy security. Critical minerals, which are required for low carbon technologies, have very poor diversity. The top producing country has more than half of the share of the extraction of rare earths, cobalt, lithium, and platinum **(Figure 14.8)**[813,814].

Global Share of Production

- ➢Lithium: Australia has half of the global share
- ➢Rare Earths: China has two thirds of the global share
- ➢Graphite: China has more than two thirds of the global share
- ➢Cobalt: Congo has more than two thirds of the global share
- ➢Platinum: South Africa has more than two thirds of the global share

Figure 14.8 The global share of the top producing country for key critical minerals. Data Source[815]: IEA

The critical minerals require a processing step following their extraction. The processing of critical minerals is even less diverse. China has a dominant share in all aspects. It has a very large share in either the extraction or processing of many critical minerals. Chinese companies have also made major investments in countries–such as Australia, Chile, Congo, and Indonesia–that produce critical minerals. China also dominates the mass manufacturing of low carbon technologies. China's domination of the entire supply chain is because of its unique conditions such as a) access to a cheap and large workforce, b) the strong influence of an autocratic government and c) access to mineral resources. Other nations are very unlikely to diminish China's domination in the foreseeable future. Such dominance of China is alarming. This situation is a major concern for the energy security of other nations.

The previous transition was driven by two critical advantages. Fossil fuels had a superior cost and convenience of energy use compared to biomass. The situation is different for the low carbon transition. Basic science dictates that the low carbon technologies will be more costly than fossil fuel technologies in the foreseeable future[816]. Also, the best-case scenario for low carbon technologies would be to achieve a similar level of convenience.

There are no market incentives to drive the low-carbon transition. Government policies will need to drive it. This is far more challenging. Why? Because policies are likely to change with local and global conditions. Examples of such conditions are political changes, a rise in energy security concerns, natural disasters, a change in public perception, and poor relations between countries.

IEA has stressed the importance of a high level of international co-operation[817]. IEA estimates that a low level of international co-operation can delay the transition by four decades. Why is this an extraordinary challenge? Because each country has its own challenges and politics, and therefore its own agenda.

The conflicting interests and priorities of the countries make it very challenging to maintain a high level of co-operation. This is especially difficult for countries that have a very different governing structure such as a democracy vs. an autocracy. The challenges related to high international co-operation are clear from the tax wars and unequal access to covid vaccines across the globe. There was inadequate co-operation about covid vaccines even though tens of millions of lives were at stake[818,819]. The international co-operation was low despite the obvious and acute importance. This is very revealing.

The energy transition at the proposed speed has many challenges that are unprecedented. The challenges are so extraordinary that success will require several miracles. Influencers who lack expertise in the energy field have caused a lot of confusion about this topic. This has led to the irrational belief in many that a fast transition is straight forward.

The reality is that the energy transition has many extraordinary challenges. This has crucial implications. If the transition is forced at an unreasonably high speed, energy costs will rise sharply because of the fierce competition for limited resources such as critical minerals and skilled workers. Key issues–such as high costs, access to land, availability of difficult technologies, the increasing not-in-my-backyard attitude, global and local politics, and energy security–will force a much slower shift to low carbon energy. The only way to avoid very large delays is to design a pathway that weighs all the challenges.

Chapter Highlights

Crucial facts are those that include all the essential information to address climate change in a sustainable manner. The crucial facts are provided below:

- Climate change is a serious problem that requires urgent attention.
- Climate change is not the only serious problem. Global society is faced with several other serious problems that require urgent attention.
- There is no climate change related cliff in the foreseeable future. The global society is not on the brink of a massive devastation because of climate change.
- The standard of living and human development index of low- and mid-income countries depends on their energy use.
- There is no perfect energy technology option. Each option has distinct advantages and drawbacks.
- The shift from fossil fuels to low carbon energy will be enormously costly.
- A low carbon world has a very high risk of severe environmental issues.
- The energy transition is enormously complex and has many extraordinary challenges.

…§§§-§-§§§…

15. Robust vs. Poor Policies

Many countries are in the early stages of developing energy policies to drive the shift to a net zero world. The support for the policies will depend on the level of public trust[820]. Public trust in the governments is low in many countries. The public trust in the United States is only 31%[821]. The average public trust in OECD group of countries is only 41%[822].

Public trust is low because of the poor track record of global governments. Poor management of the country's budget and partisan politics is a common concern in most countries. Poor budget management has caused the national debt to rise drastically in many countries[823]. Also, partisan politics often drive policies. Such policies do not address the concerns of the public. Poor policies are a recipe for low public trust.

What does this mean for energy policies?

Energy policies will have a significant impact on every person. So, policies that are not efficient or appear to be partisan will further erode the public trust. Poor energy policies will not retain support.

How we address the climate and energy problem will deeply impact every person on earth. Robust polices will lead to a bright future. Poor policies will cause immense damage. I cannot overstate the importance of the quality of policies. Policies will only be successful if they receive public support over the long term. The policies must be robust to receive such support.

What is required for robust policies? A focus on the science and realties of climate and energy. That is the road to a brighter future.

Here, I discuss the specific requirements that robust policies must satisfy and clarify how these requirements can distinguish between robust and poor policies. I also discuss a path to sustainably address the climate and energy challenge.

Requirements For Robust Policies

The crucial facts about climate and energy were discussed in the last chapter[824]. These crucial facts serve as a guide to the requirements for robust policies.

Below, I will consider the requirements that arise from each of the crucial facts.

The first crucial fact is that climate change is a serious problem. This crucial fact tells us that robust policies must show an urgent intent to address climate change.

The second crucial fact is that climate change is not the only serious problem. This tells us that robust policies must focus on the efficient use of resources. This will ensure a fair distribution of our limited resources to mitigate climate change and address other urgent problems.

The third crucial fact is that there is no credible evidence for a climate related cliff in the foreseeable future. This tells us that the end-of-the-world rhetoric must not drive policies. Robust policies must be based on the science and realities of climate and energy.

The fourth crucial fact is that the economic state of a country depends on its energy use. This tells us that robust policies must ensure that affordable energy is available to all countries.

The fifth crucial fact is that a perfect technology option does not exist. This tells us that robust policies must not be driven by passion about any technology. Robust policies must recognize that the optimal option will depend on the location, situation, existing technology mix, and the time of deployment.

The sixth crucial fact is that the energy transition will be very costly. It will increase the cost of energy. The level of increase will depend on the energy policies. Robust policies must ensure that the costs are manageable.

The seventh crucial fact is that the energy transition will very likely have a severe impact on the environment. This tells us that robust policies must focus on decreasing the side effects related to the transition.

The eighth and final crucial fact is that the energy transition has many extraordinary challenges. Robust policies must carefully address these challenges. For example, the energy transition will require unprecedented resources at an ultrafast pace. This will very likely lead to energy supply issues and high costs. Robust policies must pay special attention to these concerns. They must ensure energy is available at an affordable cost to the global society.

Robust Policies

Based on the earlier discussion, a summary of robust policies is listed below.

- The policies must urgently address climate change.
- The policies must ensure that the cost remains manageable.
- The policies must ensure the most effective use of resources.
- The policies must not be driven by emotions.
- The policies must ensure that there is no risk to global energy security.
- The policies must consider the environmental side effects.

To be consistent with the science and realties of climate and energy, robust policies must satisfy all the requirements stated above.

To summarize, energy policies are robust if they address climate change in a cost-effective, and efficient manner and address the concerns about energy security and environmental side effects.

These robust policies represent the optimal and quickest route to address climate change. Why? Because such policies are the most likely to retain public support. It will be impossible to mitigate climate change without long-term public support.

Below, I provide details about the critical aspects that robust policies should focus on.

Policies Should Be Cost-effective

Policies should focus on cost-effective options to reduce greenhouse gases. Cost effective options are those that have competitive upfront and lifetime costs within the sector. Policies that give top priority to cost-effective options will maximize the impact of the budget. The targeting of the low hanging fruit will ensure the biggest bang for the buck.

The **current** cost-effective options for reducing greenhouse gases from the power sector are listed below[825].

- Utility solar and onshore wind are cost effective until a certain level of use. The cost-effective level of solar and wind power is likely to be between 10 and 30% in most regions.
- Extending the life of nuclear power plants is very cost effective.
- Hydropower and geothermal power are cost effective when suitable locations are available.
- Biomass power is cost effective when low-cost feedstock is available.
- New nuclear power is cost-effective in certain regions.
- Short duration battery storage is cost-effective in some regions.
- Natural gas power is cost-effective in regions with low or moderate natural gas prices.

The current cost-effective options to reduce greenhouse gas emissions from the cars sector are mass transit and hybrid cars[826]. Mass transit has the least upfront and lifetime cost per passenger mile. This is because personal cars require far more vehicle miles–when compared to mass transit–for the same number of passenger miles.

A massive expansion of mass transit must be strongly incentivized over personal cars in densely populated regions. The situation is different in regions with a low population density. Mass transit by itself is not practical in such regions. Policies should encourage an expansion of mass transit, and the use of hybrid cars in regions with a low population density. Hybrid cars are cheaper than electric cars and offer more convenience.

An improvement in fuel efficiency is currently the most cost-effective option for decreasing emissions from freight trucks, ships, and aircraft. Policies should focus on increasing fuel efficiency to decrease emissions until the low carbon options become cost-effective.

Heat pumps are cost-effective and practical for homes in regions with low carbon electricity and mild climates[827]. Policies should encourage the use of heat pumps in such regions.

What about the low carbon options that are not yet cost-effective?

These technologies should be given high priority at a later stage when they become competitive. For example, the technology options required to reduce 35% of the CO_2 in a net zero world are still in the demonstration or prototype stages[828]. It will take decades for these options to become cost competitive. Governments should provide funding for R&D efforts and early adoption to make them cost competitive. Policies should not give a high priority to widely deploy these options at the present stage[829]. There will be a sharp rise in energy prices if the global society is forced to use technologies before they are cost competitive. Policies should emphasize the use of these options at a later stage when they become cost competitive.

Policies Should Be Efficient

Policies should ensure an efficient use of our resources.

One robust approach is to deploy the low carbon options in a sequence that optimizes the efficiency of greenhouse gas reduction.

Recall the wise proverb: "*Do not place the cart before the horse*". Efficient policies avoid the poor use of resources by ensuring that the cart is not placed before the horse. This is especially relevant to electrification. The ability of electrification to reduce greenhouse gases depends on the carbon intensity of the power sector. For example, electric cars reduce greenhouse gases by less than 20% in regions where the power sector has a high carbon intensity[830]. So, policies should not force a large use of electrification options until the power sector has achieved a low carbon intensity. Policies should also ensure that the electrical grid can handle the excess electricity demand from electrification. Giving a high priority to electrification before the electrical grid achieves low carbon intensity and high reliability is not efficient. It is like placing the cart before the horse.

A major shortfall is expected for skilled workers in the energy space. Policies should make it easy for the workforce to access quality training to develop the required skillsets.

Policies should focus on climate adaptation.

Mitigating the impact from climate change by decreasing greenhouse gases is a slow process. Even in the best-case scenario, the global society will face a significant impact from climate change for decades. Climate adaptation can provide quick relief from the climate impacts in the interim. The goal of climate adaptation is to make the global society more resilient to the impact from droughts, floods, heat waves, wildfires, and storms[831,832]. I will discuss a few examples of climate adaptation below.

- Drought resistant farming and maintaining adequate access to water in regions prone to droughts.
- Building levees and improving water drainage systems in regions prone to flooding.
- Improving access to cooling in regions prone to heat waves.
- Improving assessment and awareness of wildfires, improving fuel management and land use, creating buffer zones and fuel breaks, and changing building codes in regions prone to wildfires.

Climate adaptation is a very efficient tool to quickly lower the impacts from climate change. This is especially critical to decrease the suffering of the poor. Policies should provide adequate resources to climate adaptation.

Policies Should Address Concerns About Energy Security

Energy security refers to an affordable and reliable access to energy. There are two major concerns about global energy security. Policies should address both these concerns.

The first concern is about the risk of energy shortage that can result from poor assumptions. There are many extraordinary challenges associated with the proposed transition[833]. There is no evidence to suggest that global society will meet these challenges. Based on the unrealistic assumption that the global society can meet all these unprecedented challenges, it can be tempting to restrict new fossil fuel projects[834]. **This is a major concern.**

It is crucial that the policies related to new fossil fuel projects should be based on credible evidence. They should not be based on a fanciful wish or assumption. This is crucial. Why? Because restricting new fossil fuel production prematurely will cause a major shortage of energy. The energy demand of the global society will not be met and there will be global chaos. *To avoid this, policies should not restrict the supply of fossil fuel energy until the energy supply from low carbon technologies is proven to be adequate and affordable.* Instead, the focus should be on decreasing the demand for fossil fuels. We can achieve this by encouraging the use of low carbon energy[835].

The second concern is about the very poor geographical diversity associated with low carbon energy. We saw earlier how a single country, China, dominates the supply chain for the popular low carbon options. This poses a serious energy security risk to the rest of the world.

Policies should focus on improving the geographical diversity. The process will be slow because of the realities associated with the mining and processing of critical minerals. For example, the global average lead time for a new mining project is two decades. This highlights the need for adjusting the expectations about the speed of the transition.

Policies Should Address Concerns About the Environmental Side Effects

Low carbon options have a high resource intensity because of the low help from nature[836]. Consequently, they are very likely to have a severe impact on the environment when used on a massive scale. Policies should use three approaches to address this concern.

Polices should focus on using the most diverse set of low carbon options possible[837]. This approach is based on the principle that the harmful effect of a substance is less harmful when it is used in a dilute form. The use of a diverse set of options will result in a lower level of use of each technology. This will lead to a lower harmful impact because of the dilution effect. A diverse use of resources is critical to achieve a dilution effect[838]. Let us consider the example of fossil fuels as a reference point. The share of fossil fuel energy in the past decades has been about 85%. Its impact on climate change would have been markedly lower if the share had been limited to less than 40%[839].

Use of diverse options has one more key benefit. It will relax the stress related to the land and critical minerals requirements. For example, nuclear power and natural gas with CCS require far less land and critical minerals compared to solar and wind power. So, increasing the share of these technologies will lower the stress.

Policies should discourage the waste of energy and encourage options that are energy efficient[840]. This will result in lower use of resources and thereby a lower impact on the environment. For example, policies should strongly incentivize mass transit, i.e., buses, rail, and shuttles[841]. Policies should not encourage poor energy habits by giving massive incentives to personal cars.

Policies should encourage the reduction of food waste. According to United Nations, about a third of all food produced is lost or wasted[842]. The global food sector consumes 30% of the global energy supply and produces 20% of the emissions[843]. So, a drastic reduction in food waste can significantly reduce energy consumption and emissions.

Policies should ensure that there are timely regulations to limit the environmental impact from the low carbon options. Let us again consider the example of fossil fuels. If the use of carbon capture and storage had been regulated in a timely manner, the impact from fossil fuels would have been smaller. Timely regulations also give adequate time to address the practical issues related to the proposed solutions.

Poor Policies

To facilitate this discussion, I will reiterate the key elements of robust policies. Robust policies are consistent with the science and realities of climate and energy. They focus on addressing climate change in a cost-effective and efficient manner. At the same time, they address concerns about energy security and environmental side effects.

Poor policies do not satisfy one or more of these points. So, they are not consistent with the science and realities of climate and energy.

Most countries around the globe have agreed to reduce greenhouse gases with urgency[844]. But there are major concerns about the global path forward. The proposed path includes many poor policies.

Poor Policies in the Power Sector

The use of poor policies is common in the power sector. Let us consider a couple of cases below.

The first case relates to policies that promote high-cost options for wide deployment[845].

The European Union and the United States have energy policies that give a high priority to options such as rooftop solar and offshore wind[846,847]. These options are not currently cost-effective to reduce greenhouse gases. They have a much higher upfront and lifetime cost than the other options[848,849,850].

The second case relates to inefficient policies that are too focused on certain options.

Several countries have policies that are too focused on solar and wind power. These policies ignore options such as nuclear power. This is mainly because of irrational fear[851]. The policies in Germany serve as a good example.

Solar and wind power have been the main focus of German policies in the recent decades. This has greatly increased the share of solar and wind power in its electricity mix[852]. At the same time, their policies have focused on decreasing the share of nuclear power. Germany has been shutting down nuclear power plants. Dispatchable nuclear power has been replaced by intermittent solar and wind power. These policies have had undesirable effects. The use of coal power has only decreased slightly despite the large increase in renewable power[853]. This is a concern because coal power is the most polluting source of electricity.

What is the overall result of these poor policies in Germany?

Its power sector has a markedly higher intensity of greenhouse gases compared to peer countries **(Figure 15.1)**[854]. Notably, the peer countries have a much lower share of solar and wind power. The intensity of the German power sector is comparable to the global average[855]. Despite the massive emphasis on solar and wind power, Germany is far from achieving the leadership position it had hoped for.

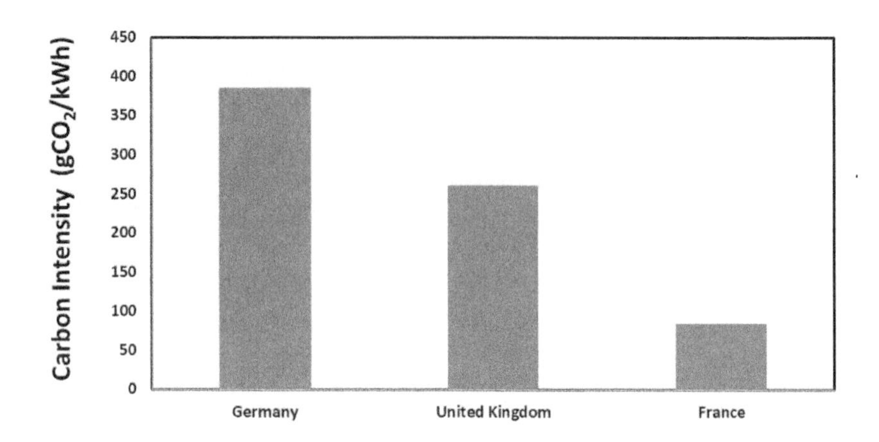

Figure 15.1 Comparison of the carbon intensity of the electrical grids in Germany, United Kingdom, and France. The carbon intensity is expressed in terms of grams of CO_2 equivalent per kWh of electricity. The data is for the year 2022. Data Source[856]: Ember Climate

The state of California is another example. It has placed a major emphasis on solar and wind power in its electricity mix since more than a decade. The policies have rapidly increased the share of solar and wind power in California from 2010 to 2020 **(Figure 15.2)**[857]. The goal was to sharply decrease the carbon intensity of its electricity generation. But that did not happen[858]. Increasing the share of solar and wind power has had only a small impact in decreasing the carbon intensity **(Figure 15.2)** [859].

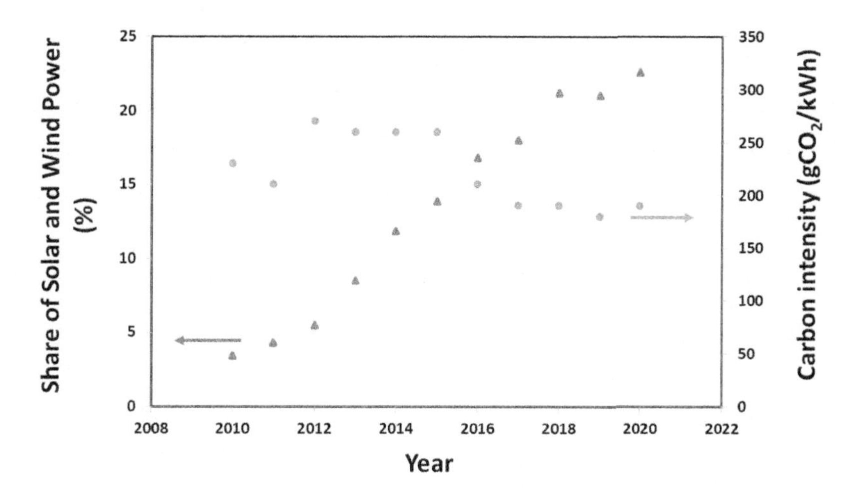

Figure 15.2 The share of solar and wind power (solid triangles) in California over the years along with the carbon intensity (solid circles) of its electrical grid. Data Source[860,861]: California energy commission and California air resources board.

There is another critical problem associated with an overemphasis on solar and wind power. A diverse energy mix is key to lower the severity of environmental impacts from the energy transition[862]. So, it is critical to include as much diversity in the low carbon options as possible. The policies must not ignore any competitive option.

Poor Policies Related to Electrification

Current policies that give high priority to electrification are not efficient in many regions.

Electrification has high upfront costs[863]. More importantly, the electrical grids are not yet ready to support efficient electrification. The electrical grids must have low carbon intensity and must be reliable to support efficient electrification. That is not the case.

Most regions of the world have an electrical grid with a moderate or high carbon intensity[864]. Electrification is not efficient to reduce greenhouse gases until a low carbon intensity is achieved. Recall, an electric car can reduce greenhouse gases by less than 20% in regions that have an electrical grid with a high carbon intensity.

Electrical grids are aging in many countries. Also, the grids are not equipped to handle large increases in solar and wind power without major upgrades[865]. There are reliability concerns even in advanced economies such as the United States[866,867].

Electrification greatly increases the demand for electricity. So, electrification can increase the stress on the grids and cause reliability problems. A massive upgrade of electrical grids will be required globally[868]. The upgrade is critical to maintain high reliability. The upgrade efforts have been slow because of the lack of smart prioritization.

The importance of poor energy policies is being increasingly recognized. For example, the North American Electric Reliability Corporation has included energy policies as a risk to energy reliability in its most recent report[869]. The report identifies the speed of the changing resource mix and aspirational policies as major risks.

The electricity demand is expected to more than double as we progress to a net zero world[870]. To avoid disruption and ensure high efficiency each region must first achieve a low carbon and reliable grid. A major push towards electrification should only occur after this. Many countries are giving a high priority to electrification before addressing the basic concerns about the electric grid. This is a classic example of placing the cart before the horse. Such policies lead to an inefficient use of resources.

Many policies related to electrification focus on electric cars. China, United States and European Union have been giving a high priority to electric cars since many years[871]. Electric cars are being touted as a solution to rapidly reduce greenhouse gases. But the impact from electric cars is sensitive to the carbon intensity of the electrical grid. Most of the global population lives in regions, where the grid has a carbon intensity that is medium or high[872]. Electric cars can only reduce the greenhouse gas emissions between 5% to 60% in such regions[873,874,875].

Apart from concerns about the electrical grid, electric cars also have many other drawbacks compared to the other options[876]. Expansion of mass transit is a vastly superior option in regions with a high population density[877]. And mass transit coupled with hybrid cars are superior in regions with a low population density. The alternatives have a lower cost for CO_2 avoidance and can reduce more CO_2 in most regions.

Poor policies ensure a poor use of our limited resources. Let us consider an example. Electric cars and battery energy storage compete for the same critical minerals. It is far more efficient to first address the grid reliability and carbon intensity. This can be partially achieved by policies that give a higher priority to utility-scale battery storage in certain regions. Also, electric cars are taking away money from other cost-efficient and efficient options. This is a problem because all options compete for the same limited budget. Such an inefficient approach makes it more difficult to address climate change.

The policies in China about electric vehicles are instructive. China has provided massive subsidies to electric cars since more than a decade[878,879].

But China has also given a top priority to coal power plants at the same time. The electricity generated by coal has increased by over 50% during the same period[880,881]. China uses more coal than the rest of the world combined[882]. China has been building a massive number of new coal power plants[883,884]. In the year 2022, the coal power capacity being built in China was six times more than in the rest of the world combined.

The enormous increase in the use of coal in China over the last decade is a concern because coal power is one of the largest sources of greenhouse gases[885]. It is also the most polluting source of electricity.

Notably, China also has some of the most aggressive policies to support solar and wind power. But the massive addition of coal power has decreased the overall effectiveness to reduce emissions.

The carbon intensity of the electrical grid in China is still high because of the large use of coal[886]. Replacing conventional cars with electric cars can reduce greenhouse gas emissions only between 5% to 40% in China[887,888,889,890]. Giving a top priority to electric cars to address climate change in China is not efficient.

China has provided tens of billions of dollars of subsidies to electric cars[891,892]. These subsidies would have been far more impactful if they had been used to decrease the use of coal power. Several times more greenhouse gas emissions could have been reduced. Why? Because low carbon power has a much lower upfront cost for reducing greenhouse gas emissions compared to electric cars[893]. Giving a top priority to electric cars in China to address climate change is a poor use of resources.

Expanding mass transit is a far more superior option than electric cars in China. A laser-like focus on expanding mass transit in China will have multiple benefits. It will decrease greenhouse gases more rapidly, at a lower cost and with fewer environmental side-effects[894].

The incentives for personal electric cars are also problematic from a road congestion viewpoint. Road congestion is already a massive problem in China[895]. This is despite a low vehicle ownership of less than 0.3 per person[896]. For comparison, the vehicle ownership in the United States is 0.9 per person. A focus on incentives for personal electric cars will increase the car ownership in China. This will radically worsen the road congestion problem. Despite many issues, China continues to massively subsidize electric cars.

General Issues Related to Poor Policies

Poor policies do not consider the extraordinary challenges associated with the proposed shift to low carbon energy. The speed of the shift is not consistent with the resource availability. They encourage a premature halt to new oil and gas production based on an irrational optimism about the speed of the shift to low carbon energy. This will cause frequent disruptions in the global energy supply. The resulting backlash will make it very difficult to address climate change.

There is another major risk that is often ignored. Consider a country or a state that attempts the transition at a very fast pace. Such a country or state will be forced to use options that are not cost-effective or efficient. This will increase the energy costs and thereby the cost of all goods and services. High energy costs are not conducive to manufacturing industries. The industries will be forced to move to other regions. A large fraction of the population will be significantly affected by the higher costs. The overall economic harm will be immense. Moreover, the local benefits related to lower climate impacts will be insignificant[897]. This will cause a massive backlash and climate policies will lose public support. **There is massive downside but no upside for any region to have policies that are very costly or inefficient.**

What Does Global Investment Data Show?

Data on global investments show that the energy sector is beset by poor policies. Such policies are leading to inefficient global investments. Recent data is revealing[898].

- There is a critical need to upgrade the global grids. This should get the highest priority. Yet the global investment in grids is low compared to that of renewable power. The investments are very low relative to what is required. For example, the recent Inflation Reduction Act and Bipartisan Infrastructure Law of the United States include an investment in grids that is still short by over a hundred billion dollars[899,900,901].
- Electric cars do not deserve a high priority. Nuclear power and battery storage, which are robust options to improve energy reliability, deserve a much higher priority. Yet, the global investment in electric vehicles has been more than the investment in nuclear power and battery storage combined in the last three years[902].

Implications of Poor Policies

A budget discipline is important considering that the global debt is skyrocketing. The global government debt has increased from $22 trillion in 2000 to an incredible $92 trillion in 2022[903]. The total global debt which includes borrowing by governments, businesses and households is over $300 trillion[904]. Poor energy policies will lead to a further large increase in the global debts. This will make it unbearable for the future generations.

Poor policies waste our limited resources on options that are not effective. This leads to a lower resource availability for efficient options.

Massive subsidies are being used to support poor energy options. Poor policies are being supported under the assumption that such policies are the fastest approach to address climate change. But this is a wrong assumption. Poor policies cannot be sustained over the required duration. Why? Because they are neither cost-effective nor efficient[905]. Poor policies will also lead to energy security problems, global chaos, and increase environmental issues. The is expected because poor policies are not consistent with the science and realties of climate and energy.

The impact from poor policies will not stay hidden for long[906]. The climate benefits from poor policies will be low. The public will face escalations in energy cost and a host of other problems. For example, poor policies will take away resources from crucial actions related to climate adaptation and addressing poverty. The resulting public backlash will delay the energy transition by decades.

Using robust policies is the fastest approach to address climate change. The use of poor policies to increase the speed will cause harm and long delays. It is understandable that a very fast speed is desirable. But such a desire is not consistent with science and realities.

Consider the example of a patient who is suffering from cancer. A robust treatment is available. It involves chemotherapy for several weeks followed by a surgery. Trying to speed up the process by forcing all the chemotherapy in a few days will not result in a better outcome. A faster cure is desirable. But moving away from a robust treatment and using a poor treatment will do far more harm than good. It will not increase the speed of a good outcome. Instead, it will greatly increase the side effects and will result in a poor outcome.

The Big Picture About Policies

The global path forward must include only robust policies to ensure a successful energy transition. But avoiding poor policies is not easy. Many poor policies are being driven by misinformation that is very convincing. There are two pieces of misinformation that are popular[907].

The first is that climate change will have a catastrophic impact on the human society soon.

The impact from climate change is expected to become more severe with every increment of temperature rise. But the data and scientific consensus does not support the narrative that cataclysmic events are likely in the foreseeable future[908]. Unfortunately, the glut of misinformation has pushed many into believing this false narrative.

The second piece of the misinformation is that the energy transition is not very difficult.

This is far from the truth. The transition has many extraordinary challenges because of the unprecedented nature of the change.

Let us review some challenges associated with the proposed energy transition[909].

- Resources such as critical minerals, land, and skilled workers will be required in extraordinary amounts. There is no evidence to indicate that we can access such resources at the desired speed. It is very likely that there will be a frequent imbalance in the supply and demand for these resources. Past data shows that an imbalance can cause a sharp increase in the cost of the resources.
- An unprecedented level of innovation will be required. Why? Because the technologies required to decarbonize the most difficult energy applications are still in the demonstration or prototype stages. These technologies are required to reduce 35% of the CO_2 in a net zero world[910,911]. Also, most of the commercially available low-carbon technologies do not have a comparable cost or convenience relative to fossil fuel technologies. For example, electric cars require much longer wait times for recharging and have a shorter travel shorter range compared to conventional vehicles. The uptake of the popular low-carbon technologies to this point has been mostly driven by massive subsidies. It is not clear when many of the low-carbon technologies will become competitive.
- Global energy security is also a major concern. The geographical diversity is very poor for critical minerals and the manufacturing of low carbon technologies. China dominates most of the supply chain.
- The cost of the energy transition is estimated to be in the hundreds of trillions of dollars.
- Severe environmental side effects are very likely because of the gigantic resource intensity of the energy transition.
- An unprecedented level of global collaboration will be required to avoid a major delay. The reality is that countries are moving in the opposite direction. Disputes, trade wars and other isolationist policies are on the rise globally.
- Historical data is also very instructive to discuss the challenges. The prior transition was ten times smaller and yet took a hundred years. So, the proposed speed is a major challenge. The average lead time for a new mining project for some key critical minerals is twenty years. Innovation in the energy field also requires a long time. Relatively easy technologies such as solar and wind

power required between 30 to 60 years to reach a significant market share. It gets worse. The technologies that are still in the demonstration phase are also the most challenging. They address the most difficult energy applications such as air travel, ships, and high temperature industrial processes. Several decades will be required for these technologies to become competitive.

The global society has never attempted anything of this magnitude and complexity. Not even close. Examples such as vaccines for COVID-19, space missions, digital revolution are not relevant. They required trivial resources and were far less complex compared to the energy transition. The challenges faced by **all our prior successes combined are trivial** compared to the extraordinary challenges of the energy transition.

Major disruptions in energy supply and cost escalations are very likely if policies do not respect the challenges. When this happens there will be backlash against the policies and the global governments will further lose credibility. Overall, the efforts to mitigate climate change will be greatly hindered if the challenges are not given due attention.

How do we ensure success? By focusing on the science and realities of climate and energy.

The global focus should only be on robust policies. We looked at the attributes of robust policies earlier. Robust policies address climate change in a cost-effective and efficient manner. They also address the concerns related to global energy security and environmental side effects at the same time. Robust policies are consistent with science and respect the practical constraints.

We will first discuss a generic framework of robust policies that applies at a global level. Next, we will discuss how to apply this framework to individual countries. Finally, we will examine the implications for the global society.

Generic or Global Framework of Policies

This framework applies to all countries and consists of a group of policies with a common goal of sustainably addressing climate change. The policies target the following:

- Give the highest priority to upgrade the transmission infrastructure. Why? To ensure a reliable and efficient supply of electricity.

- Give top priority to all the cost-effective options that can reduce greenhouse gases from the power sector. Why? For two reasons. The decarbonization of the power sector is a critical step in elevating the impact of electrification. The use of all cost-effective options equals to a higher impact on emission reductions. This also means that there will be more diversity. The dilution effect from the diversity is key to decrease the severity of the environmental impact from the low carbon options. The use of a diverse set of options will also decrease the stress related to land and critical mineral requirements.
- Give the highest priority to decrease emissions from coal power. Why? Because coal power alone contributes to 20% of the global greenhouse gas emissions. It is the most polluting fossil fuel based on greenhouse gas emissions and air pollution.
- Focus on reducing the demand for fossil fuels but not their supply. Why? To avoid disruptions in energy supply that could result from a premature shortage of oil and gas.
- Focus on reducing methane leakage. Why? Because it is a cost-effective and efficient approach to reduce emissions.
- Give a high priority to electrification only after the electrical grid has achieved a low carbon intensity and high reliability. Why? To ensure an efficient use of our limited resources and to avoid disruptions in electricity supply.
- Give top priority to the massive expansion of mass transit. Give high priority to ride sharing and the use of hybrid cars. Why? Because these options are cost-effective and efficient[912,913].
- Give top priority to improve the fuel efficiency of ships, air travel, and long-distance freight trucks. Why? Because low carbon options are far from being competitive for these applications.
- Give top priority to options that can improve energy efficiency in a cost-effective manner. Why? To decrease energy consumption while avoiding a sharp rise in energy cost.
- Provide support for R&D and early adoption for low carbon options that are currently not cost-effective or efficient. Why? Because this is an effective approach to increase their competitiveness. We should deploy these options with high priority after they become competitive.

- Adjust the speed of the transition based on the resource availability and other constraints. Why? Because forcing a speed that is too fast will cause a sharp rise in energy prices. Price is a function of supply and demand. A speed that is too fast will not allow the supply to keep up with the demand. The resulting short supply of key resources such as critical minerals will have a large impact on the energy prices.
- Give top priority to energy security. Why? Because adequate access to energy is critical for a good standard of living.
- Give a top priority to encourage the workforce to develop the required skill sets. Why? To avoid the predicted shortage of skilled workers in this space.
- Introduce timely regulations. Why? To limit the environmental impact from the low carbon technologies.
- Give high priority to reduce energy and food waste. Why? This is a very efficient approach to address climate change.
- Give top priority to climate adaptation. Why? To quickly lower the severity of the impact from climate change.

Country Specific Policies

Earlier, I discussed a generic framework of policies. This framework can be used to develop more detailed policies for each country.

Countries differ from each other in terms of the carbon intensity of their grid, fuel prices, population density, energy use, and economic state. The generic framework can be used to develop policies for any country by considering its specific features. Here, we will use United States as an example. Specifically, we will apply the generic framework to United States by considering its specific features.

The electrical grid of the United States has a medium carbon intensity. Its carbon intensity is below the global average[914]. But a large reduction in emissions will still be required to achieve a low carbon intensity[915].

Coal power is the major contributor to the emissions in the power sector. It contributes to 55% of the greenhouse gas emissions despite having a less than 20% share of the total electricity[916,917]. Decreasing the emissions from coal power deserves a high level of urgency.

But there are challenges. PJM, the largest regional grid operator in the country, has documented the challenges in a recent report[918]. PJM has expressed concerns about the risk to the grid reliability. The concerns arise from two issues. The first issue is that dispatchable resources such as coal power are being replaced mainly with intermittent resources such as solar and wind power. The second issue is that the electricity demand is growing because of electrification. The impact from these issues is expected to increase each year. This will make it difficult to meet the year-round electricity demand in the coming years. The risk to the reliability will be especially high during periods of low sunshine and calm winds[919].

These challenges are common to all the electrical grids in the United States[920]. The challenges can be mitigated by using robust policies.

Policies should give the highest priority to upgrade the aging electricity transmission infrastructure.

The power sector also deserves urgent attention. Policies should focus on replacing coal power with all the available cost-effective options to reduce emissions. Along with utility solar and onshore wind, several other options should also be given high priority. Natural gas power is a very cost-effective option in the United States because of low natural gas prices[921]. It has a very low upfront cost and low lifetime cost[922]. It is a dispatchable resource and has the highest flexibility to balance supply and demand. Replacing coal power with natural gas power can reduce emissions by 50%[923]. A further 70% reduction can be obtained by including a carbon capture and storage (CCS) component. This component can be added when the CCS technology becomes cost-effective[924].

Options such as hydropower and geothermal power should also be given top priority in locations where they are cost-effective. Extending the life of nuclear power plants deserves the topmost priority. New nuclear power projects and short duration energy storage should be given priority at suitable locations.

The use of this diverse mix of power options in the United States has many benefits. The benefits include a reliable electricity supply, optimized costs, and a lower environmental impact.

What about electrification?

Electrification options have a high upfront cost[925]. Electrification is not efficient to reduce greenhouse gas emissions until the power sector achieves a low carbon intensity. For example, consider the impact from electric cars and heat pumps. The impact on emissions is markedly lower for a grid with medium carbon intensity compared to one with low carbon electricity[926,927]. The United States grid has a medium carbon intensity. So, the impact of electrification is much lower than what could be with a low carbon grid. Moreover, the reliability of the grid is a major concern as discussed by the PJM report.

Electrification markedly increases the use of electricity[928]. This can increase the stress on the grid[929]. So, policies should give a high priority to electrification only after the power sector has achieved a low carbon intensity and high reliability. This is especially true for applications where superior alternatives are available.

The expansion of mass transit is superior to subsidizing electric cars in the United States and elsewhere. Mass transit is more cost-efficient to reduce greenhouse gas emissions. Mass transit can also reduce the emissions more efficiently. Currently, electric cars can reduce emissions by 40 to 65%[930,931]. Buses, vans or mini-buses, and rail can reduce the emissions by 60 to 80%[932,933]. The upfront and lifetime cost is much lower per passenger mile for mass transit. This is because personal cars require far more vehicle miles when compared to mass transit for the same number of passenger miles. Consider the comparison of a personal car with a bus. A bus can carry over 50 passengers. Many more personal cars are required to travel the same number of passenger miles as a bus. The use of personal cars leads to many times more vehicle miles. More vehicle miles translate to more maintenance and operating expenses. More vehicles also mean a higher upfront cost. So, personal cars are less cost-effective than mass transit. Personal cars include electric, conventional, and hybrid cars.

The need for fewer vehicle miles in case of mass transit leads to several other benefits[934]. It leads to a lower use of energy and resources[935]. It decreases traffic congestion and the risk of accidents. So, mass transit is also superior to decrease the environmental side effects and resolving traffic issues. Policies should give a foremost priority to expanding mass transit in high-density population centers. A rapid expansion of buses or transit vans is very practical[936]. These options do not require special infrastructure and can be deployed rapidly.

Many regions in the United States have a low population density. Mass transit by itself is not a practical option in such regions. A combination of mass transit and hybrid cars is a superior option in regions with a low population density. Hybrid cars can only reduce emissions modestly, by 25 to 30%[937]. But they have a much lower upfront cost and are more convenient than electric cars. An average person is far more likely to embrace hybrid vehicles because of the superior convenience and lower upfront cost[938]. The shift to hybrid cars will be several times faster. Also, hybrid cars do not increase electricity consumption[939]. So, hybrid cars are far more practical to reduce greenhouse gas emissions in the short term compared to electric cars [940]. Policies should give top priority to expanding mass transit and hybrid cars in regions with a low population density.

The United States has a high energy use per person[941]. It uses four times more energy than the global average and two times more than the European Union. Policies should focus on lowering the energy use. The key is to lower the energy use in a practical manner. This can be achieved by a focus on options that minimize the waste of energy, and materials[942]. Example of such options for the transportation sector are expanding mass transit, and increasing the fuel efficiency of freight trucks, ships, and air travel. Other examples are using air conditioners judiciously, insulating homes, promoting the multiple use of items, and reducing food waste.

Heat pumps can lower the energy use in homes in certain states. Heat pumps are cost-effective and efficient in states that have a low carbon grid and mild climate[943,944,945]. Policies should give a high priority to heat pumps in such states[946].

Its high energy use makes energy security a major concern for the United States. The supply chains for the low carbon energy options have extremely poor diversity. China dominates the supply chains, while the United States has a very small share[947]. The United States policies should focus on increasing its share in the supply chains of low carbon technologies. It should also decrease its dependence on countries that have an autocracy. Democratic countries have a high potential for political misalignment with such countries. An increase in self-sufficiency is very desirable. Policies that focus on lowering the energy use can also help with this[948].

The energy transition will take several decades because of the extraordinary challenges. So, oil and gas will be a critical component of the energy mix in the foreseeable future. Any realistic estimate agrees with this[949,950].

To ensure global energy security, countries such as United States and Canada, who have large reserves, must continue to be major producers. These democracies are politically stable and more reliable. Also, they are most likely to minimize the pollutants during the production of oil and gas. Examples of other top producers of oil and gas are Russia, Saudi Arabia, Iran, and Venezuela.

What will happen if oil and gas production is discouraged in countries such as the United States? The global society will need to increasingly rely on less reliable and more polluting producers. So, policies must not discourage oil and gas production in the United States.

Policies should focus on creating a workforce that has the required skill sets to deploy operate and maintain the massive projects that will be required for the low carbon energy world.

Policies should give a foremost priority to climate adaptation. Climate mitigation and adaptation must go hand-in-hand for the best possible outcome.

United States has been the major contributor to global warming. So, it should play a leading role in the energy transition. But this must be done efficiently.

I will discuss a few examples.

Many low carbon technologies are not yet competitive. United States has the resources to play a leading role in making these competitive. Policies should support R&D efforts and the early adoption of such technologies. These technologies should be given top priority for wide deployment after they become competitive.

Policies should support the transfer of technology know-how to developing countries to reduce emissions. Also, partnerships should be established to lower the use of coal in developing countries. This will have a large impact on reducing emissions. Such policies are far superior to providing large subsidies to options that are not yet cost-effective or efficient in the United States. Examples of less effective options include rooftop solar, offshore wind and electric cars.

Subsidies should be eliminated for these options. Climate enthusiasts should drive the voluntary use of options such as rooftop solar and electric cars until they become competitive[951]. Climate enthusiasts believe that climate change is the most serious problem. Many climate enthusiasts in the United States are affluent and should be able to afford the additional costs. So, climate enthusiasts should become the cheer leaders of such technologies.

United States should also play a leading role in ensuring that the energy transition does not cause major environmental side effects. It should introduce timely regulations to limit the environmental impact from the low carbon options.

Policies should support climate adaptation with the topmost priority. Many regions in the United States are already experiencing a severe impact from climate change. The fastest approach to lower the impact is via strategic climate adaptation efforts.

I will summarize the policies discussion.

In the first phase, the policies in the United States should focus on following.

- Upgrade the transmission infrastructure. Decrease carbon intensity of the power sector by using cost-effective options.
- Expand mass transit, and increase fuel efficiency of cars, ships, freight trucks, and air travel to decrease emissions from the transportation sector.
- Lower the energy use by minimizing waste of energy, goods, and food.
- Encourage all options that can reduce energy use in a cost-effective and efficient manner.
- Greatly increase its share in the supply chain of low carbon energy technologies.
- Support other countries to decrease their emissions in a cost-effective manner.
- Massively increase R&D efforts to improve the low carbon technologies that are not competitive. Make it easy for the workforce to access the training required to develop the required skill sets.
- Support timely regulations to limit the environmental side effects from the transition.
- Support climate adaptation efforts.

Success in the first phase will efficiently set up the next phases. The other low carbon options should be given top priority in the next phases as they become competitive.

The specific strategies for the different phases will be determined by the four crucial aspects: cost-effectiveness, efficiency, energy security and environmental side-effects.

This phased approach, which is based on prioritization, is consistent with the science and realities of climate and energy.

What about other countries?

Countries differ in their features. For example, United States has a medium intensity grid, while Canada has a low intensity gird. The specific policies in each phase will be different for different countries. But there are common points that apply to the policies in all the countries. These points serve as a guide for policies to robustly reduce greenhouse gas emissions in each phase.

For wide deployment, all countries should focus only on the competitive options in each phase. A competitive option has a comparable cost and performance to the incumbent[952]. R&D and innovation will increase the competitiveness of the other options over time. These options should be widely deployed in the later phases once they become competitive.

Policies in each phase should ensure that the energy supply continues to be secure and affordable. All countries should design policies to minimize energy waste. The policies should reduce dependence on unreliable countries for energy supply. Countries must also ensure that energy costs do not escalate.

Countries such as India that have an energy use far below the poverty-buster levels must be especially careful. Why? Because the energy transition is going to markedly increase energy costs. Poor policies will lead to very high energy costs. This will force a lower energy use and will prevent improvements in the standard of living of Indian citizens. This challenge can be mitigated via the use of robust policies.

Climate adaptation is crucial to all countries. Policies in each country should dedicate adequate resources to climate adaptation. This will ensure a fast relief to its citizens and increase the credibility of government efforts.

Policies should ensure that environmental side effects from the transition are lessened. The use of diverse options, i.e., dilution, is critical for decreasing the side-effects. So, countries should use all the competitive options that are available. The focus should not be only on solar and wind if other competitive options are also available. For example, nuclear power should be considered as an important option in many countries.

Implications

Systematic prioritization is the key component of a robust framework of policies. A robust framework is consistent with the science and realities of climate and energy. It is the fastest possible path to mitigate climate change because it addresses all crucial issues about climate and energy.

A framework of policies designed to meet a highly aspirational target will downplay the extraordinary challenges of the energy transition. A systematic prioritization of energy options will not be a foundation of such a path. A path without systematic prioritization cannot speed up the transition. Such a framework will include many poor policies. It will not address all the crucial issues about science and energy. The use of such a framework will not be sustainable. The global society will reject it, and this will cause major delays.

Our best approach is to embrace a policy framework that is sustainable. Only a policy framework that addresses all crucial issues about climate and energy is sustainable. Any framework that includes poor policies will cause substantial harm to the global society. It will lose public support and will cause decades of delay.

Adopting a robust policy framework–i.e., focusing only on robust policies and avoiding poor policies–will be crucial. But that is very challenging. Why? Because of the rampant misinformation about energy. This misinformation is the reason poor policies continue to be proposed. We will review how to avoid misinformation in the next chapter.

Chapter Highlights

- Robust energy policies are those consistent with the science and realities of climate and energy.
- To be robust, the low carbon energy policies must satisfy four crucial requirements. The policies must a) be cost-effective, b) be efficient, c) address energy security concerns and d) minimize environmental side-effects.
- A systematic prioritization of energy options is crucial to a framework of robust energy policies. The shift to low carbon energy must occur in phases. The four crucial requirements should be used to prioritize between the different energy options in each phase.
- Only robust policies will be able to retain long term public support. Consequently, the use of a framework that only includes robust policies is the fastest path to mitigate climate change.
- Poor energy policies do not satisfy one or more of the four crucial requirements. Such policies fail to correctly prioritize between the energy options.
- The current global policy framework is a mix of robust and poor energy policies.
- A framework of policies that is based on a highly aspirational goal cannot be consistent with the science and realities of climate change and energy. Such a policy framework will not allow a smart prioritization of energy options. So, it will contain many poor policies. Such a policy framework will soon lose public support and cause decades of delay.
- Climate adaptation is crucial. Adequate resources should be dedicated to climate adaptation. This will ensure a fast relief from the impact of climate change and increase the credibility of the climate related efforts.

...§§§-§-§§§...

16. How to Avoid the Misinformation Trap?

There are many examples of poor polices being deployed across the globe. These poor policies are being mainly driven by the rampant misinformation about climate and energy.

The misinformation crisis is also causing several other problems. It is increasing the hatred between activists and skeptics and increasing anxiety in much of the young population. Also, it is decreasing the confidence in scientists and institutions.

It is crucial to avoid the misinformation trap. Critical thinking can assist with this goal. This quote from George Washington is pertinent[953]. "*Truth will ultimately prevail where pains is taken to bring it to light.*"

The material provided in this chapter can help to decide whether the information is trustworthy. Specifically, it can help to address two critical questions about the information under consideration.

- The first question relates to the source of the information. Is the source credible?
- The second question is about the nature of the information. Is the information of good quality?

This chapter focusses on how to evaluate the credibility of the source, the quality of the information and how to avoid misinformation related to research papers.

How to Evaluate the Credibility of the Source?

We are being flooded with information about climate and energy. Mainstream and social media, books, and scientific reports all contribute to the information blitz.

This discussion applies to both primary and secondary sources of information. A primary source is the creator or original source of the information. A scientist who conducts research to generate the information is an example of a primary source. A secondary source mainly shares the information that is generated by somebody else.

Examples of secondary sources can be news channels, social media posts, books, and blogs.

The credibility of the source of the information is a good indicator of its trustworthiness. A source with high credibility is far less likely to share misinformation and vice versa. The credibility of the source can be evaluated by asking one simple question.

Does the source consistently display a balanced view about climate and energy?

A positive response to this question indicates a high credibility. A negative response is a red flag. This increases the risk of misinformation from the source.

I have discussed earlier why certain people have an excessive fear about climate and energy. Being a captive of excessive fear is a major concern because it leads to extreme views. Those with extreme views have a narrow view. Their view is mainly driven by emotions. This makes it very difficult for them to provide balanced information.

I will reiterate the two extreme views about climate and energy.

Skeptics believe that climate change is not a serious problem. They downplay the impact of climate change. They believe that there is no urgent need to decrease greenhouse gas emissions. They only focus on the positives of fossil fuels and the negatives of renewable energy.

Activists believe that climate change is by far the most serious problem facing mankind. Many believe that the human society is on the brink of destruction because of climate change. They want an immediate shift to low carbon energy. They believe that a very rapid shift is only a matter of political will. They do not think that there are any extraordinary challenges. They only focus on the negatives of fossil fuels and the positives of renewable energy and electrification.

The two extreme views do not consider **all these** crucial facts about climate and energy.

- Climate change is a serious problem and must be addressed urgently.
- The proposed energy transition has several extraordinary challenges such as a rapid access to unprecedented resources and major technology gaps.

- No energy technology is perfect.
- Fossil fuels have been extensively used because of their unique advantages.
- A very rapid shift can endanger the global energy security.

A balanced view is consistent with all the crucial facts about climate and energy. It accepts that a) greenhouse gas emissions must be urgently reduced by rapidly increasing the use of low carbon energy, b) the speed of the transition will be slower than desired because of the extraordinary challenges, c) there is no perfect low carbon energy option and d) fossil fuels will be important to the global energy security in the foreseeable future.

A balanced view considers the science and realities about climate and energy. The extreme views do not. That is why a source with extreme views in not likely to provide balanced information. A balanced view is a critical requirement for a high credibility of a source.

I will summarize the questions that can be used to evaluate the credibility of the source.

Does the source exaggerate or downplay the impact of climate change? Does the source see only the positives of some low carbon options and only the negatives of others? Does the source only praise or vilify fossil fuels? Does the source downplay the challenges associated with the energy transition? Does the source NOT consistently demonstrate a balanced view?

If the response is a yes to any of these questions, then the source is not credible and is likely to be a channel for misinformation. We should not believe the information provided by the source without further verification. This is especially true for the information that is based on major assumptions. This is because assumptions are a natural entry point for bias in the predictions involving complex systems such as climate and energy. So, such information must be independently verified from a credible source.

If there is no way to independently verify the information, it is best to ignore it. This is an effective way to avoid the misinformation trap because it is practical and robust.

How to Evaluate the Quality of the Information?

The misinformation risk can be minimized by avoiding poor quality information[954]. The quality of the information is poor if a) it omits key facts, b) it uses assumptions that cannot be validated, c) it is not broadly applicable or d) it is not generated by an expert.

The information is of poor quality if falls into either of the categories. Below, details are provided about each category.

Information that omits key facts.

We will recall some examples that showcase this category.

Example 1. Skeptics tout that climate has always been changing since the beginning of time. They use this to claim that climate change is not a concern. But they omit a key fact. The climate has been changing at a very rapid speed over the past several decades[955]. This is unlike how climate has changed in the distant past. The unusually rapid speed is a cause for serious concern.

Example 2. A recent study reported on the global deaths that occurred over the past two decades because of suboptimal temperatures[956]. The study showed that the global deaths related to hot temperatures have increased. Activists use this data to advertise that this is an alarming consequence of climate change. But they omit the other critical results of the study. The study revealed that the global deaths related to cold temperatures have decreased[957]. The cold related deaths were almost ten times higher than the heat related deaths. So, the overall effect was that deaths because of suboptimal temperatures have declined over the past two decades. Activists ignore this key positive aspect of the paper. They only discuss the negative aspects, which is very misleading. Such misuse of data from research papers is common.

Example 3. Activists mainly focus on the advantages of solar and wind power. They ignore the challenges. This supports their preferred narrative of very rapidly shifting to solar and wind power.

Example 4. Skeptics mainly focus on the challenges of solar and wind power. They ignore the advantages. This supports their preferred narrative of avoiding the use of solar and wind power.

The examples show that the use of selective facts leads to the wrong conclusions. This type of misinformation can be very convincing. Typically, it is difficult to know when key facts are being omitted from an article, report, or social media post or a book.

How to address this challenge?

An effective solution is to find and consider the counter narrative. All the key facts can often be found by considering both the narrative and its counter narrative.

Information that is based on poor assumptions.

Information can be broadly classified into two types.

The first type is mainly based on quality observations. It does not require major assumptions. If some assumptions are required, they have a sound basis. Moreover, the assumptions do not have a large impact on the quality of the information. There is transparency and accountability associated with this type of information.

Examples of this category are historical data about weather, climate impacts, GDP, population, poverty levels, energy use, and energy production. Governmental and intergovernmental agencies are credible sources for this type of information. I have extensively used this type of information in this book from organizations such as IEA, IPCC, World Bank, and other official agencies of the United States, European Union, and United Nations.

The second type of information is based on major assumptions that do not have a sound basis. This type of information is of poor quality. We will examine two examples.

Example 1. Skeptics assume that the impacts of climate change can be managed via adaptation alone. They use this assumption to conclude that it is not necessary to reduce greenhouse gases urgently. A claim like this should be backed by concrete evidence. But such evidence does not exist. Climate change has a significant impact on the entire ecosystem. It is well understood that the impact will increase markedly as the warming increases. There is also an increased risk from very high impact events. The probability of such events is small at the lower levels of warming. But the probability increases with increasing warming. So, there is no sound basis to assume that adaptation alone will be effective in managing the long-term impact of climate change. It would be very reckless to bet our future on such a weak assumption.

Example 2. Activists believe that a very rapid shift to low carbon energy is not very difficult. They assume that the global society is well equipped to overcome all the challenges related to a very rapid transition. They use this assumption to conclude that no new oil and gas projects are required[958]. This conclusion will have serious consequences if it is wrong. The resulting energy shortages will cause chaos in the global society. The backlash will markedly delay the efforts to mitigate climate change. So, robust evidence is essential to support their assumption. But such evidence does not exist. There is no evidence to suggest that the global society can address the extraordinary challenges of a very rapid shift. Let us recall some specific challenges and the ability of the global society to meet them[959]. A very rapid access to an unprecedented level of resources such as critical minerals, skilled labor and land will be required. This is no evidence to suggest that the global society can get such access. The technologies required to address the most difficult 35% of the emissions are still not available. An unprecedented level of success in energy technology innovation will be required to have these technologies available for wide use in few decades. We have not achieved such success in the past, not even close. Also, a very high level of international co-operation will be required. The required level of collaboration has not been seen before. To make it worse, countries have been moving in the opposite direction in recent years. The trade wars, disputes and other isolationist policies are on the rise. Activists assume that the global society will address these and the other extraordinary challenges. There is no evidence to support this assumption[960]. *In fact, there is strong evidence to the contrary.* Based on the historical records, it is very likely that the global society will not be able to meet any of these challenges. The global society is in a wholly unchartered territory. It would be very dangerous to bet our future on poor assumptions.

It is easy to identify this category of poor information. We need to ask two questions. Are the assumptions important to the main message? Are the assumptions mainly speculative?

If the answer to the questions is a yes, then the information should be avoided. *It is important to ignore this type of speculative information irrespective of the source. Such information is of poor quality even if it is from governmental or intergovernmental organizations.* I have avoided using speculation-based information in this book.

Information that is not broadly applicable.

We will start this discussion with two examples.

Example 1. A few countries already have a very high share of renewable power. Activists use this information to suggest that all countries should be able to shift to renewable power very quickly. But this is only possible for the few countries because they satisfy certain unique conditions[961]. They either a) have an adequate number of suitable sites for hydropower and geothermal power[962] or b) can export and import electricity to balance the supply and demand gap caused by a large share of solar and wind power[963]. Most countries cannot not satisfy such unique conditions. So, this information is not broadly applicable. It only applies to a few select countries.

Example 2. Many believe that nuclear power is not an option. They use the high-profile accidents and a few projects with large cost and time overruns to support their claim. This information is not representative and so is not broadly applicable. Over the decades, nuclear power has showed a very good safety profile when compared to other power technologies[964]. Also, many projects have not had major cost or time overruns. That is the reason the share of nuclear power has been significant over the decades[965,966].

To avoid being misled by this type of poor information, it is best to ignore the arguments that are based on unique conditions or limited data.

Information that is not generated by an expert.

Climate and energy are both extremely complex topics. Only experts can generate quality information (data) or make major advances in these fields.

A climate expert must have knowledge about climate monitoring tools and the various processes that can influence the climate. An expert must also understand the complex interactions within the climate system. The researchers from universities and other academic institutions have the skills, experience, and motivation to develop expertise.

Energy experts must have experience with developing energy technologies, evaluating a wide range of energy technologies and with major energy projects[967]. The energy industry provides unique access to the relevant expertise[968]. So, most energy experts have either a current or past affiliation with the energy industry. It takes a minimum of fifteen years of relevant work in the energy industry to gain such experience[969].

The above information can be used to identify climate and energy experts. Any major claim related to the climate that is based on information generated by someone without an expertise in climate science should be ignored. Any major claim related to the energy transition that is based on information generated by someone who does not have the relevant energy expertise should be ignored.

Let us consider my example. I have over twenty years of experience in the energy field, most of it being in the energy industry. This includes experience with evaluating and developing energy technologies and major energy projects. I have relevant energy expertise but not relevant climate expertise. Consequently, I have only used climate information that has been created by climate experts in this book.

Overall, I have taken special care to focus only on good quality information to draw conclusions. I have ignored information that is based on poor assumptions, or which is not broadly applicable.

Key Issues Around Research Papers

There are four topics of concern related to research papers. Each can lead to misinformation about climate and energy.

Misrepresentation of Research Papers

Activists and skeptics often use mainstream and social media to misrepresent the information from research papers. Crucial facts are often omitted from the research papers.

If they want to present a pessimistic narrative about a topic, they mainly focus on the worst-case scenarios reported in the papers. They omit or downplay discussions of the other scenarios. If they want to present an optimistic narrative, they mainly focus on the best-case scenarios.

How to avoid being misled? By reading the summary of the original research paper. A summary or abstract of most research papers can be accessed via the internet and is free[970]. The summary of a research paper highlights all the crucial facts. It only takes minutes to browse through the summary.

Errors In Research Papers

Scientists around the globe document their research in scientific journals. Most of the journals use a peer review process. Typically, two or three peer researchers from the same field serve as reviewers. This process is an efficient approach to regulate the quality of the research papers. But it does not guarantee research papers that are error-free.

Millions of papers are published each year. Each paper is a result of months of research from the authors. It is not practical for the peer reviewers to verify the accuracy of the mathematical models or experiments. This would take several weeks at a minimum. A peer reviewer can usually only spend some hours on each paper. So, the peer reviewers cannot check the paper in-depth for errors. They can only check if the method used in the paper is reasonable, and whether the data analysis supports the conclusions.

Despite the best intentions of the authors and the peer reviewers, there is a potential for errors. The potential for errors is higher in complex fields such as climate science and energy.

I will discuss an example. The journals Nature and Science recently reported that a few research papers have overstated the climate impacts because of their use of poor-quality climate models[971,972]. Fortunately, this has not caused any problems for the recent IPCC report. The discussion about future climate impacts in the IPCC report is based on superior quality climate models[973]. The IPCC report is far more reliable because the discussions are drawn from the efforts of multiple research groups.

There are many sources of errors in research papers. I will discuss two common sources. One source is the use of inaccurate data in the analysis. Inaccurate data is mostly a result of poor data collection methods. The second source of errors in research papers is the use of data that is not representative.

The peer review process typically cannot identify papers with data related errors. Why? Because the peer reviewers can dedicate only a limited time to each paper. A proper data review requires several weeks at the minimum. Also, the peer reviewers often do not have in-depth expertise about the papers they are reviewing because of the numerous sub-specializations in complex fields. So, they cannot ensure that the data used in the papers is error free. This is especially true in complex fields such as climate and energy. In practice, the reviewers have to assume that that the data used in the papers is free from errors. That is why such papers can pass the peer review process.

How to avoid such misinformation?

By ignoring isolated papers that make sensational claims. Most sensational claims are a result of poor data or assumptions used in the papers. Sensational claims are those that differ drastically from scientific views that are based on years of research from multiple research groups. This is the most efficient approach to avoid getting misinformed by errors in the research papers[974]. It is best to consider the expert consensus for important matters.

Speculation in Research Papers

Most research involves three steps. Data is collected in the first step. The data is analyzed in the second step. The analysis is used to draw conclusions in the final step. The conclusions from such research papers have a high level of accuracy if the data and the analysis are robust.

But what about research papers that attempt to predict future changes in a system?

Robust data is not available to support such predictions in most cases.

The lack of robust data is an obvious concern. For example, consider a research paper that attempts to predict the economic state of a country in the year 2100. There are many unknowns that can impact the conclusions. Will the country be impacted by natural disasters or a health crisis?[975] Will it be involved in wars? Will it have political stability and competent leadership?

To predict the far future state in complex systems is very challenging. This is because there are several unknowns which can play an important role. This is especially true when trying to predict any major change that may occur far out in the future. An example would be a paper that attempts to predict the timeframe when a poor country will become one of the richest in the world. The role of unknowns becomes even more important when trying to make predictions about major changes.

Research papers typically use assumptions to deal with the unknowns. Due to lack of relevant data, such assumptions are speculative and are not robust. This is a major problem because a poor assumption results in a poor prediction. The situation is far worse for papers that attempt to make predictions about major changes. The conclusions from such papers are even more sensitive to the speculative assumptions. So, such research papers have a very poor accuracy or a very large uncertainty. The conclusions from such papers do not help to make robust decisions.

The above discussion directly applies to speculative research papers about climate and energy. I will discuss specific examples below.

The scientific community has a very little understanding about the very high impact events that could occur because of climate change. This is because our climate system is very complex, and relevant data is not available. The consensus amongst climate experts is that very high impact events are unlikely in the foreseeable future. The risk from such events is low according to the consensus. However, a few research papers disagree. These research papers sound an alarm about such events based on highly speculative assumptions. These papers are attempting to predict a major change, although there are many key unknowns about the climate system. These papers are forced to use highly speculative assumptions and that is why the predictions from such papers are not reliable.

Speculation is also a major problem with research papers that discuss the impacts of the energy transition. Research papers have provided very optimistic estimates for a rapid transition. These papers are forced to make many assumptions. This is because there are many unknowns related to transforming the global energy system. Some assumptions are direct while some are tacit. For example, many papers tacitly assume that the global society will be able to address the extraordinary challenges related to a rapid energy transition[976]. There is no evidence to support the critical assumptions used in these papers[977]. That is a major problem because the conclusions are sensitive to the assumptions. The use of speculative assumptions leads to conclusions that lack credibility. Such papers do not provide information that can be used to make robust decisions.

The most efficient approach to avoid being misled is to ignore research papers that are based on speculative assumptions. Recall, speculative assumptions are those that do not have a sound basis.

Breakthroughs vs High-impact breakthroughs

Research papers and start-ups often report breakthroughs related to novel energy technologies. These breakthroughs are, in most cases, not a precursor to commercial success[978]. There is a major difference between a breakthrough and high-impact breakthrough. A breakthrough is any major scientific or technical advance. But a high-impact breakthrough is a special type of advance that enables commercial success. High-impact breakthroughs are extremely rare in the energy field. A breakthrough is far more common. Most breakthroughs are not high-impact breakthroughs.

It is critical to avoid mistaking a breakthrough to a high-impact breakthrough. Unfortunately, this is a common error in mainstream and social media. This type of error leads to an undeserved optimism about the success of new energy technologies. How to avoid such confusion? By avoiding judgment until the data about commercial success is available for a new technology. Such data is only available after a new technology has achieved a significant share of the global market.

How to Avoid Misinformation about Policies

Politicians make all sorts of claims about energy policies. A common claim is that their proposed energy policies will create jobs and improve the economy. Such claims are not very meaningful. They do not provide any information about whether the policies are robust.

An efficient way to avoid misinformation about policies is by asking the following questions.

- Do the policies give top priority to the cost-effective options to reduce greenhouse gases?
- Do the policies give high importance to the efficiency of resource use?
- Do the policies place an emphasis on short and long-term energy security?
- Do the policies focus on lowering the overall environmental impact?

We can also get more specific with the questions.

Do the policies focus on ensuring a low carbon and reliable grid before electrification? Do the policies give a foremost priority to mass transit? Do the policies focus on reducing the demand for fossil fuels as opposed to their supply?

Robust policies require a positive response to these questions. If the answer is a "no" to these questions, then the policies are not robust.

Chapter Highlights

- The most effective way to avoid misinformation is by evaluating the credibility of the source of information. If the source does not routinely provide balanced information about climate and energy, it should be a red flag. Information from such a source should only be considered if it can be independently verified.
- Another effective way to avoid misinformation is by avoiding poor quality information. The misinformation related to poor quality can be avoided by a) considering both the narrative and counter narrative about the topic in question, b) rejecting any argument that is based on speculative assumptions, c) rejecting any argument that is based on unique conditions or very limited data and d) ignoring information that is created by non-experts.
- Misinformation resulting from the misrepresentation of research papers in social media can be avoided by reading the summary (abstract) of the original research paper.
- Misinformation related to the errors in research papers can be avoided by ignoring sensational claims from isolated research papers.
- Unreliable information can be avoided by ignoring research papers that use speculative assumptions to predict major changes.
- Misinformation related to poor policies can be avoided by asking the following questions. Are the policies cost-effective? Are they efficient? Do they address energy security concerns? Do they address the concerns about environmental side-effects? Policies that do not satisfy these crucial requirements are not robust.

…§§§-§-§§§…

17. Final Thoughts

Activists and skeptics are engaged in a massive misinformation war. The warring groups are leaving no stone unturned. They are using mainstream media, social media, research papers, books and more to win the war.

A key goal of the activists and the skeptics is to gain public support for their views about climate and energy. Such support is the key to align energy policies with their views.

This war has increased in intensity in recent years. Who is winning the war? What are the implications? How can we ensure a better future?

We will discuss the answers to these critical questions in this final chapter.

Who is winning the war?

The war is still raging. But it appears that the activists are winning. There are several reasons for this.

One key reason is that the misinformation from skeptics is less believable to a neutral observer.

Most of the misinformation from skeptics is along the following lines. "The earth is not warming. CO_2 is not causing warming. More CO_2 is good. Climate change is a hoax. Solar and wind power are very poor options."

Such misinformation is not consistent with well-advertised facts. The neutral observer can sense a red flag.

The misinformation from activists is *also not consistent* with the science and realities about climate and energy. *But it is more believable.*

Most of the misinformation from activists is along the following lines. "The impact of climate related disasters has increasing very rapidly in recent times. A rise in temperature above 1.5°C will be devastating. The energy transition is not very difficult. Solar and wind power are very cheap. A shift to low carbon energy will lower the global cost of energy. We continue to use fossil fuels mainly because of greedy corporations. We have all the technologies needed for a very rapid transition. The energy transition will eliminate our environmental problems."

Such misinformation is more believable because it is based on a distortion of facts. Substantial knowledge about climate and energy is required to recognize that this is misinformation. This makes it difficult for a neutral observer to sense any problem.

Another reason is that many more people from academia identify as activists than as skeptics. Teachers and professors have a high credibility and a greater influence.

Also, media outlets have placed a much greater emphasis on their coverage of natural disasters in recent years. This gives a false impression that the impact from disasters has been increasing very rapidly[979].

The impact of such influence can be seen from a recent survey[980]. The survey targeted people aged between 16 to 25 from ten representative countries. 75% of those surveyed said that the future is frightening because of climate change. **Over 55% believed that humanity is doomed**.

Clearly, activists are winning the climate misinformation war. Moreover, they are also winning the energy misinformation war. An average person does not have access to robust information on energy issues. Most energy experts–who can provide such information–work in the energy industry. Personnel from the energy industry rarely publish research papers, articles, or books[981]. Publishing is mainly of interest in academia. Unfortunately, most academics who publish papers about the energy transition do not have relevant energy expertise. They make many poor assumptions, which leads to research papers that display an optimism about the energy transition that is not realistic[982]. It gets worse. The poor information provided by these papers is touted in mainstream and social media.

There are massive new opportunities in low carbon energy because of the enormous injection of funds by the global governments. This has led to a massive growth in start-up companies related to low carbon energy. These companies are adding to the confusion. It is very common for the start-up companies to *vastly over-promise to get funding*. Mainstream and social media repeat the exaggerated claims made by the start-ups. The public is not aware of these issues and believes that the claims are true. Thus, many start-ups are contributing to the unrealistic optimism about the energy transition.

The result is that a neutral observer is mainly exposed to unrealistic information about the energy transition.

Strong financial interests can also lead to a misinformation bias. Being employed by a specific industry or access to large investments can both lead to bias.

Which industry employs more people? The low-carbon industry employs *over 50%* of global energy workforce and is expected to grow its lead[983].

What about investments? The investments in the low carbon industry have also been markedly higher in recent years[984]. For example, the investment was *65% higher* in the low-carbon industry compared to the fossil fuel industry in 2022.

The above discussion shows why activists are in a better position to win the war and bring most of the global population to their side.

Potential Implications

What would happen if the global policies were influenced by the misinformation from either the activists or skeptics?

The implications are obvious if the misinformation from skeptics was to prevail. This will lead to global policies that continue the unabated use of fossil fuels. There will be no urgency to reduce greenhouse gases. This will greatly increase the impact from climate change over the next few decades. The resulting delay will greatly increase the difficulty and cost of the future climate mitigation efforts. It will also increase the risk of very high impact events. *Overall, it will cause a massive setback to the global society.*

But the above scenario is not likely. The activists seem to be winning the misinformation war. What will happen if this continues, i.e., the activists win the misinformation war?

This more likely scenario also has terrible implications. The misinformation will cause extreme anxiety about climate change in the global population. It will lead to an unrealistic optimism about the energy transition. Overall, this will lead to global policies that are cost-ineffective and use resources inefficiently. It will also lead to policies that prematurely decrease the supply of fossil fuels[985]. Such policies will disrupt the energy supply and cause a sharp rise in energy costs. Moreover, the greenhouse gas reductions will be small because of the poor efficiency of the policies. This small reduction will come at a high cost. The low income and middle class across the globe will be severely impacted[986,987].

The impact of the poor policies will be felt in a decade or so. The pain from the impact will cause a backlash against the energy policies. It will also lead to a loss in trust in the scientists and institutions[988]. This will make it very difficult to address climate change. Overall, this scenario will also cause a massive setback to the global society.

The misinformation from skeptics and activists will both greatly increase the difficulty of addressing climate change. This is ironic considering that activists and skeptics have opposing views about the impact of climate change.

A Brighter Future

There is only one clear path to a better future. The global society *must ignore the extreme views* of activists and skeptics[989]. This path will move us past the misinformation about climate and energy. This path will lead the majority away from extreme views and will enable robust global policies.

Why are extreme views so harmful? Because such views do not leave adequate room for critical thinking. This makes it is very difficult for activists and skeptics to avoid the misinformation trap.

People who do not have extreme views can use critical thinking to avoid misinformation. I will summarize my earlier discussion on this topic.

- The most effective approach to avoid misinformation is to evaluate the credibility of the source. If the source has not demonstrated a balanced view about climate and energy, it is a red flag. The broad conclusions from such a source cannot be trusted.
- One other approach is to consider the narrative and counter narrative for each claim. This helps to avoid misinformation that is based on the omission of crucial facts.
- A very effective approach is to ignore information that is based on speculative assumptions[990]. Assumptions that do not have a sound basis are speculative. They are commonly used to make

predictions about very high impact climate events and the cost and speed of the energy transition. Such predictions increase confusion and misinformation. Information based on speculative assumptions must be ignored from all sources. This includes experts and governmental or intergovernmental organizations.

- Use of outlier data is a common type of misinformation. There is a simple check for this type. If broad conclusions are drawn from unique conditions and narrow data, it should raise a red flag. It is best to ignore such conclusions.
- Isolated peer reviewed research papers can also be a source of misinformation. It is best to ignore sensational claims from isolated peer review papers. Instead, the focus should be on the collective knowledge from multiple research groups.
- Climate and energy are extremely complex topics. If the original source of the information is not an expert, it is likely to be of poor quality. Misinformation can be avoided by ignoring information that is created by non-experts.

The good news is that the global society can easily stop the misinformation crisis.

Only a small fraction of the global population has extreme views. Most people are not activists or skeptics. They do not have very strong views about climate and energy. This majority is the key to stop the misinformation crisis.

How? By becoming more involved.

The best path is to become a climate pragmatist. I define climate pragmatists as those who respect the science and realties about climate and energy. They are driven by facts, not by excessive fear. *They are optimistic, but their optimism is grounded by reality.*

We need the majority of the population to become climate pragmatists. If the global majority plays a larger role as pragmatists, they can end the misinformation crisis. They can strongly influence the global policies. **Such a majority can force robust global policies.** The politicians will do what the majority demands. Robust policies will ensure that we restrict the rise in temperature to the lowest that is practically possible. This is the most efficient approach to minimize the climate and environmental impact and ensure a robust and affordable supply of energy.

Together as climate pragmatists, we can pave the path to the best possible future of energy and the environment. **A brighter future is in our hands.** Hopefully, we can work together and make it a reality.

Chapter Highlights

- The misinformation about climate and energy from activists is more convincing than that from skeptics. Consequently, activists are winning the misinformation war.
- The misinformation from skeptics and activists will both result in poor policies. So irrespective of which extreme side wins the misinformation war, it will be major loss for the global society.
- Fortunately, there is a promising alternative. If majority of the population decides to play a larger role as climate pragmatists, the global society can move past the misinformation crisis. Climate pragmatists are those who are passionate about the well-being of our society and do not have extreme views about climate change and energy.
- If we embrace the role of climate pragmatists, we can ensure robust energy policies across the globe. A brighter future is in our hands!

…§§§-§-§§§…

Book Review

I would be delighted to hear from you!

If you found value in this book, I kindly ask you to share your experience with others. Your feedback helps to spread the message about how to avoid misinformation about climate and energy.

Please take a minute to leave your review on Amazon or your favorite book retailer.

…§§§-§-§§§…

Glossary & Abbreviations

1 trillion = 1000 billion; **1 billion** = 1000 million; **1 million** = 1000,000

Ton: specifically used in this book as substitute for *metric ton* (i.e. herein, 1 Ton = 1000 Kg)

1 billion ton = 1 gigaton = 10^{12} Kg

Watt (W): Unit of electrical power (e.g. a typical incandescent light bulb: 60-80 W)
1 MW = 1000 kW; **1 kW** = 1000 W

Watt hour (Wh): The electrical energy unit of measure equal to one watt of power supplied to, or taken from, an electric circuit steadily for one hour. Example: an average U.S. home uses ~30,000 Wh (30 kWh) per day.

GW: Giga watts or billion watts; 1 GW = 10^9 W

TWh: Terrawatt hours; 1 TWh = 1 trillion Wh

Tons of oil equivalent: unit of energy; defined as amount of energy released from burning 1 ton of crude oil.

Crude oil: A mixture of hydrocarbons that exists in liquid phase in natural underground reservoirs.

ppm: part per million (10^{-6})

Climate resilience: ability to prepare for, recover from and adapt to detrimental impacts from climate change.

Environmental impact: For this book, this term includes a collective impact to health, safety and the environment.

Natural gas: A gaseous mixture of hydrocarbon compounds, the primary one being methane.

Ozone: Inorganic molecule with chemical formula O_3.

Plastics: Polymer (large number of similar chemical units bonded together) compounds with high molecular mass.

Economy of scale: The principle that larger production facilities have lower costs per unit produced than smaller facilities.

Subsidies: A financial contribution by any government or local public body that confers a benefit. This is the definition from the World Trade Organization (WTO)
LCOE: Levelized cost of electricity

GDP: Gross domestic product

IEA: International Energy Agency

EIA: United States Energy Information Administration

WMO: World Meteorological Organization

UN: United Nations

UNEP: United Nations Environment Programme

UN: United Nations

IPCC: International Panel on Climate Change

IMF: International Monetary Fund

CCS: Carbon Capture and Storage

About the Author

Tushar Choudhary has twenty-five years of R&D experience in addressing environmental issues in the energy field. His experience covers a wide spectrum from fundamental research to technology development & commercialization to technology analysis. Prior to retiring from the energy industry, he served in a dual role as the Director of Technology Analysis & Advancement and Sr. Principal Scientist at a multinational energy company.

He has been granted 40 U.S. & International patents and has published over 50 research papers in reputed scientific journals. He has given keynote lectures at several International Symposiums and has received numerous prestigious awards within and external to the energy industry. Some of his external awards include the Oklahoma Chemist of the Year Award (American Chemical Society), Global Indus Technovator Award (MIT), and Southwest Industrial Innovation Award (American Chemical Society). He was ranked among the world's top 2% scientists and engineers based on the career impact of his publications (Stanford University database). He is a featured expert in the climate related documentary "Close the Divide".

Tushar received his Ph.D. in Physical Chemistry from Texas A&M University, College Station; following which he joined the energy industry as a R&D scientist. One of his greatest joys during his time in the energy industry was receiving and sharing knowledge related to all aspects of technology innovation. Following retirement, his goal is to continue sharing knowledge by writing books in areas that he is passionate about.

Also by this Author: Critical Comparison of Low Carbon Technologies; Climate and Energy Decoded

Author Website: https://www.tushar-choudhary.com

References and Notes

---

[1] My dad, Dr. Vasant Choudhary, is a renowned Indian scientist with over 40 years' experience working in the field of energy. He has over 400 research papers and over 40 U.S. patents. He is the topmost cited scientist in India in the field of physical chemistry. https://fellows.ias.ac.in/profile/v/FL1993012 https://insaindia.res.in/old_website/detail.php?id=P03-1332 https://elsevier.digitalcommonsdata.com/datasets/btchxktzyw/6

[2] Merriam Webster. Definition of war. Entry 2a and b. https://www.merriam-webster.com/dictionary/war

[3] Climate change is the long-term shift in temperatures and weather patterns. United Nations. Climate action. https://www.un.org/en/climatechange/what-is-climate-change

[4] For this book, activists and skeptics are those whose views are not consistent with the science and realities of climate and energy.

[5] Fossil fuels include coal, natural gas, and fuel products from crude oil.

[6] Quote Investigator. Truth is the first casualty in war. https://quoteinvestigator.com/2020/04/11/casualty/

[7] Quote Investigator. https://quoteinvestigator.com/2014/07/13/truth/

[8] I share many numbers in the book. I round these for simplification. So, the numbers are not exact. But they are close enough to convey an accurate message.

[9] Several views of skeptics and activists are not consistent with the science and realities of climate and energy. But not all their views are inconsistent. I want to note that there are several exceptions. Activists and skeptics have a wide range of involvement with misinformation. Some are highly involved while some less.

[10] Center for Disease Control and Prevention (CDC). What is polio? https://www.cdc.gov/polio/what-is-polio/index.htm

[11] USA Today. Good genes? How can Trump eat a lot of fast food, exercise little, and be healthy? https://www.usatoday.com/story/news/politics/2018/01/16/good-genes-how-can-trump-eat-lot-fast-food-exercise-little-and-healthy/1036518001/

[12] CDC. Benefits of physical activity. https://www.cdc.gov/physicalactivity/basics/pa-health/

[13] World health organization (WHO). Healthy Diet. Key facts. https://www.who.int/news-room/fact-sheets/detail/healthy-diet

[14] United Nations. Climate action. https://www.un.org/en/climatechange/what-is-climate-change

[15] NOAA Earth system research laboratories. Education files: Natural Climate Change. https://www.esrl.noaa.gov/gmd/education/info_activities/pdfs/TBI_natural_climate_change.pdf

[16] British geological survey. What causes the earth's climate to change? https://www.bgs.ac.uk/discovering-geology/climate-change/what-causes-the-earths-climate-to-change/

[17] U.S. Environmental Protection Agency (US EPA). Causes of climate change. https://www.epa.gov/climatechange-science/causes-climate-change

[18] Energy Institute. 2023 statistical review of world energy. https://www.energyinst.org/statistical-review

[19] Wikiquote. Micel de Montaigne. https://en.wikiquote.org/wiki/Michel_de_Montaigne

[20] NASA. What is the difference between weather and climate. https://www.nasa.gov/mission_pages/noaa-n/climate/climate_weather.html

[21] IPCC: AR6. Climate Change 2021: The Physical Science Basis. Figure SPM1.a: Change in global surface temperature relative to 1850-1900. https://www.ipcc.ch/report/ar6/wg1/downloads/report/IPCC_AR6_WGI_SPM_final.pdf

[22] NASA Global climate change. Global Temperature. https://climate.nasa.gov/vital-signs/global-temperature/

[23] NASA GISS. Surface temperature analysis. FAQs. https://data.giss.nasa.gov/gistemp/faq/#q201

[24] NASA Global climate change. Global Temperature. https://climate.nasa.gov/vital-signs/global-temperature/ GISS: Goddard Institute of Space Studies

[25] NASA GISS Surface temperature analysis. Uncertainty quantification. https://data.giss.nasa.gov/gistemp/uncertainty/

[26] NASA GISS Surface temperature analysis. Frequently asked questions. https://data.giss.nasa.gov/gistemp/faq/#q101

[27] NASA Global climate change. The raw truth on raw temperature records. https://climate.nasa.gov/ask-nasa-climate/3071/the-raw-truth-on-global-temperature-records/

[28] NASA GISS Surface temperature analysis. Uncertainty quantification. https://data.giss.nasa.gov/gistemp/uncertainty/

[29] NASA Global climate change. The raw truth on raw temperature records. https://climate.nasa.gov/ask-nasa-climate/3071/the-raw-truth-on-global-temperature-records/

[30] NASA GISS: Why so many global temperature records. https://www.giss.nasa.gov/research/features/201501_gistemp/

[31] NASA GISS: Why so many global temperature records. https://www.giss.nasa.gov/research/features/201501_gistemp/

[32] NASA Global Climate change. Graphic temperature vs solar activity. https://climate.nasa.gov/climate_resources/189/graphic-temperature-vs-solar-activity/

[33] NASA Global Climate change. Graphic temperature vs solar activity. https://climate.nasa.gov/climate_resources/189/graphic-temperature-vs-solar-activity/

[34] NASA global climate change. What is the greenhouse effect? https://climate.nasa.gov/faq/19/what-is-the-greenhouse-effect/

[35] U.S. Department of Energy (DOE): DOE explains the carbon cycle. https://www.energy.gov/science/doe-explainsthe-carbon-cycle

[36] US EPA: Climate change indicators- Greenhouse gases. https://www.epa.gov/climate-indicators/greenhouse-gases

[37] American Chemical Society. Myth: Its water vapor not the CO_2. https://www.acs.org/content/acs/en/climatescience/climatesciencenarratives/its-water-vapor-not-the-CO2.html

[38] Such an effect is described as positive feedback, i.e., a response that increases the temperature.

[39] This is another way of looking at this issue. Removal of the long-lived gases contributed by the human society will decrease the earth's temperature. This will decrease the water vapor content, which will further decrease the earth's temperature.

[40] It is important to be cautious about mathematical models. These models are approximations of a system under specific conditions. If used incorrectly they can provide misleading information.

[41] NOAA Climate.gov. Climate models. https://www.climate.gov/maps-data/primer/climate-models

[42] IPCC: AR6. Climate Change 2021: The Physical Science Basis. Figure SPM1.b: Change in global surface temperature relative to 1850-1900. https://www.ipcc.ch/report/ar6/wg1/downloads/report/IPCC_AR6_WGI_SPM_final.pdf

[43] NOAA Climate.gov. Climate change: Atmospheric carbon dioxide. https://www.climate.gov/news-features/understanding-climate/climate-change-atmospheric-carbon-dioxide The CO_2 content of the atmosphere is 420 ppm currently.

[44] World health organization (WHO). Legal BAC limits by country. https://apps.who.int/gho/data/view.main.54600

[45] National Safety Council. https://www.nsc.org/getmedia/9f523dba-b7ec-4c66-9f7c-f360e4ea45dd/low-alcohol-risk.pdf.aspx Centers for disease control and prevention (CDC). Impaired driving. https://www.cdc.gov/transportationsafety/impaired_driving/states.html

[46] U.S. EPA. Learn about heat islands. https://www.epa.gov/heatislands/learn-about-heat-islands

[47] NASA. Global climate change. Can you explain the urban heat island effect? https://climate.nasa.gov/faq/44/can-you-explain-the-urban-heat-island-effect/

[48] Earth system science data. Global carbon budget 2020. Vol. 12, Page 3269, Year 2020. https://essd.copernicus.org/articles/12/3269/2020/

[49] IPCC Special Report: Global warming of 1.5 °C. Chapter 2. Figure 2.3 https://www.ipcc.ch/sr15/chapter/chapter-2/

[50] NASA. Carbon dioxide fertilization green earth: Study funds. https://www.nasa.gov/feature/goddard/2016/carbon-dioxide-fertilization-greening-earth

[51] NOAA. Science on a sphere. Ocean-atmosphere CO_2 exchange. https://sos.noaa.gov/catalog/datasets/ocean-atmosphere-co2-exchange/

[52] NASA Global climate change. Graphic: the relentless rise of CO_2. Data: Data: Luthi, D., et al. 2008; Etheridge, D.M., et al. 2010; Vostok ice core data/J.R. Petit et al.; NOAA Mauna Loa CO_2 record. https://climate.nasa.gov/climate_resources/24/graphic-the-relentless-rise-of-carbon-dioxide/

[53] Cambridge Dictionary. https://dictionary.cambridge.org/us/dictionary/english/consensus

[54] IPCC Reports. https://www.ipcc.ch/reports/

[55] NASA Global Climate Change: Scientific consensus- Earth's climate is warming. https://climate.nasa.gov/scientific-consensus/

[56] Google Scholar. https://scholar.google.com

[57] Statement from the Commonwealth Academy. https://rsc-src.ca/sites/default/files/Commonwealth%20Academies%20Consensus%20Statement%20on%20Climate%20Change%20-%2012%20March%202018%20-%202.pdf

[58] Statement from the National Academies. https://sites.nationalacademies.org/sites/climate/index.htm

[59] Statement from the American Association for Advancement of Science. https://whatweknow.aaas.org/get-the-facts/

[60] Statement from the American Meteorological society. https://www.ametsoc.org/index.cfm/ams/about-ams/ams-statements/statements-of-the-ams-in-force/climate-change1/

[61] Statement from the Joint Science Academies, https://www.nationalacademies.org/our-work/joint-science-academies-statements-on-global-issues

[62] Statement from the Geology Society of America. https://www.geosociety.org/gsa/positions/position10.aspx

[63] Statement from the Advancing Earth and Space Science (AGU). https://www.agu.org/Share-and-

Advocate/Share/Policymakers/Position-Statements/Position_Climate

[64] Statement from the American Chemical Society. https://www.acs.org/content/acs/en/policy/publicpolicies/sustainability/globalclimatechange.html

[65] List of ~200 supporting scientific organizations. http://www.opr.ca.gov/facts/list-of-scientific-organizations.html

[66] NOAA Climate.gov. Climate models. https://www.climate.gov/maps-data/climate-data-primer/predicting-climate/climate-models

[67] NOAA Climate.gov. Climate Q&A. https://www.climate.gov/news-features/climate-qa/why-should-i-trust-scientists'-climate-projections-50-or-100-years-now-when

[68] IPCC Special Report: Global warming of 1.5 °C. Chapter 2. Figure 2.3 https://www.ipcc.ch/sr15/chapter/chapter-2/

[69] European Environment Agency. Trends in atmospheric concentrations of CO_2, CH_4 and N_2O. https://www.eea.europa.eu/data-and-maps/daviz/atmospheric-concentration-of-carbon-dioxide-5#tab-chart_5_filters=%7B%22rowFilters%22%3A%7B%7D%3B%22columnFilters%22%3A%7B%22pre_config_polutant%22%3A%5B%22CH4%20(ppb)%22%5D%7D%7D

[70] NASA. Global climate change. Core questions- An introduction to ice cores. https://climate.nasa.gov/news/2616/core-questions-an-introduction-to-ice-cores/

[71] J. Hays, J. Imbrie, N. Shackleton. Science, 194, 1121 (1976). https://science.sciencemag.org/content/194/4270/1121

[72] NASA Global climate change. Graphic: the relentless rise of CO_2. Data: Data: Luthi, D., et al. 2008; Etheridge, D.M., et al. 2010; Vostok ice core data/J.R. Petit et al.; NOAA Mauna Loa CO_2 record. https://climate.nasa.gov/climate_resources/24/graphic-the-relentless-rise-of-carbon-dioxide/

[73] Mayo Clinic. Hypothermia. https://www.mayoclinic.org/diseases-conditions/hypothermia/symptoms-causes/syc-20352682#:~:text=Overview,95%20F%20(35%20C).

[74] Cleveland Clinic. Fever. What are the possible complications or risk of not treating a fever? https://my.clevelandclinic.org/health/symptoms/10880-fever

[75] World Meteorological Organization. Our mandate- Observations. https://public.wmo.int/en/our-mandate/what-we-do/observations

[76] World Meteorological Organization. Our mandate- Observations. https://public.wmo.int/en/our-mandate/what-we-do/observations The figure shows the current number and type of monitoring tools.

[77] IPCC Reports. https://www.ipcc.ch/reports/

[78] NASA Global Climate change. https://climate.nasa.gov/evidence/

[79] Each country has several scientific institutions–Universities and National Laboratories. For example, U.S. has over 400 Ph.D. (doctorate) granting institutions. https://ncses.nsf.gov/pubs/nsf21308/data-tables

[80] Use google scholar. Search term: Climate Change. See affiliations of the papers published over the decades. https://scholar.google.com/scholar?hl=en&as_sdt=0%2C44&q=climate+change&btnG=

[81] Biodiversity. Chapter 2 by Paul Ehrlich. The cause of diversity losses and consequences. https://www.ncbi.nlm.nih.gov/books/NBK219310/

[82] CDC: Overweight and obesity. https://www.cdc.gov/obesity/index.html

[83] British geological survey. What causes the earth's climate to change? https://www.bgs.ac.uk/discovering-geology/climate-change/what-causes-the-earths-climate-to-change/

[84] NOAA Earth system research laboratories. Education files: Natural Climate Change. https://www.esrl.noaa.gov/gmd/education/info_activities/pdfs/TBI_natural_climate_change.pdf

[85] IPCC Sixth Assessment Report (AR6). Frequently asked questions. The earth's temperature has varied before. How is the current warming different? https://www.ipcc.ch/report/ar6/wg1/downloads/faqs/IPCC_AR6_WGI_FAQ_Chapter_02.pdf

[86] European Environment Agency. Ocean Acidification. https://www.eea.europa.eu/ims/ocean-acidification

[87] IPCC Sixth Assessment Report (AR6). The Physical Science Basis. Summary (A.2.4) https://www.ipcc.ch/assessment-report/ar6/

[88] NASA Global Climate change. Effects. https://climate.nasa.gov/effects/

[89] U.S. Environmental Protection Agency (EPA). Climate change Indicators- heat waves. https://www.epa.gov/climate-indicators/climate-change-indicators-heat-waves

[90] IPCC Reports. https://www.ipcc.ch/reports/

[91] World Meteorological Organization Report # 1267 (2021): WMO Atlas of mortality and economic losses from weather, water, and climate extremes, 1970-2019. https://library.wmo.int/index.php?lvl=notice_display&id=21930#.YvFKgi-B3b0

[92] NASA Global climate change. Global Temperature. https://climate.nasa.gov/vital-signs/global-temperature/

[93] World Meteorological Organization Report # 1267 (2021): WMO Atlas of mortality and economic losses from weather, water, and climate extremes, 1970-2019. https://library.wmo.int/index.php?lvl=notice_display&id=21930#.YvFKgi-B3b0

[94] The values are for a period of ten years. The average impact during the last decade was 180 million affected per year and $140 billion per year of losses.

[95] World Meteorological Organization Report # 1267 (2021): WMO Atlas of mortality and economic losses from weather, water, and climate extremes, 1970-2019. https://library.wmo.int/index.php?lvl=notice_display&id=21930#.YvFKgi-B3b0

[96] Our World in Data. Natural Disasters. https://ourworldindata.org/natural-disasters

[97] IPCC Report: Climate Change 2021, Physical Basis. Contribution of working group I to the sixth assessment report. https://www.ipcc.ch/report/ar6/wg1/#FullReport

[98] European Commission. Adaption to Climate change. https://ec.europa.eu/clima/policies/adaptation_en

[99] United Nations Environment Programme. Five ways countries can adapt to climate crisis. https://www.unep.org/news-and-stories/story/5-ways-countries-can-adapt-climate-crisis

[100] IPCC Third Assessment Report (TAR), 2007 – The climate system – An overview. https://www.ipcc.ch/site/assets/uploads/2018/03/TAR-01.pdf

[101] IPCC reports and the tens of thousands of references therein. https://www.ipcc.ch/reports/

[102] IPCC Sixth Assessment Report (AR6). https://www.ipcc.ch/assessment-report/ar6/

[103] IPCC: Intergovernmental Panel on Climate Change. It was created in 1988 by the World Meteorological organization (WMO) and United Nations Environment Programme (UNEP).

[104] Steve Koonin. Unsettled: What climate science tells us, What it doesn't, Why it matters. April (2021).

[105] IPCC Sixth Assessment Report (AR6). https://www.ipcc.ch/assessment-report/ar6/

[106] Emergency events database (EM-DAT). The International Disasters Database. https://www.emdat.be I specifically report the data from Our World in Data. It uses the raw data from EM-DAT and normalizes for the change in population and GDP across the decades. The normalization provides a more accurate picture. Why? Because both population and GDP have grown enormously over the decades.

[107] Our World in Data. Natural Disasters. https://ourworldindata.org/natural-disasters

[108] United Nations. https://unstats.un.org/unsd/environment/Meteo_disasters.htm European Environment Agency. https://www.eea.europa.eu/data-and-maps/data/external/emergency-events-database-em-dat Academic papers. Google Scholar. Search string "EM-DAT". https://scholar.google.com/scholar?hl=en&as_sdt=0%2C44&q=EM-DAT&btnG=

[109] World Meteorological Organization Report # 1267 (2021): WMO Atlas of mortality and economic losses from weather, water, and climate extremes, 1970-2019. https://library.wmo.int/index.php?lvl=notice_display&id=21930#.YvFKgi-B3b0

[110] World in Data provides information per capita (per 100,000 people). This allows for an accurate comparison. It accounts for the large increase in population across the decades.

[111] Our World in Data. Natural Disasters. https://ourworldindata.org/natural-disasters The basic data about the total number of affected people is from EM-DAT. The data set normalizes for the large increase in population size across the decades. Why? Because the number of people affected also very strongly depends on the population size. If the effect of the population size is not accounted for, we will incorrectly attribute the effect of the population to climate change.

[112] Our World in Data. https://ourworldindata.org/natural-disasters# This site uses the raw data from EM-DAT and normalizes for the change in population size. Emergency events database (EM-DAT). The International Disasters Database. https://www.emdat.be UN World population aspects. https://population.un.org/wpp/ The data is only for climate related disasters. The impact from disasters that are not influenced by climate change such as volcanic eruptions and earthquakes is not included.

[113] The people affected are discussed in terms of total number of people affected per 100,000 of the total population. It is crucial to present the data in this manner to account for the large increase in population size over the decades.

[114] Our World in Data. Natural Disasters. https://ourworldindata.org/natural-disasters The basic data about the economic damage from the disasters is from EM-DAT. The data set normalizes for the large increase in GDP across the decades.

[115] Our World in Data. https://ourworldindata.org/natural-disasters# This site uses the raw data from EM-DAT and normalizes for the change in GDP. Emergency events database (EM-DAT). The International Disasters Database. https://www.emdat.be Data from GDP is from World Bank. https://data.worldbank.org

[116] The Lancet: Planetary health. Volume 5, Issue 7, 415-425, 2021. https://www.thelancet.com/journals/lanplh/article/PIIS2542-5196(21)00081-4/fulltext

[117] The paper actually discusses excess death ratios. The excess death ratio is the ratio of the excess deaths to all the deaths in a year. For simplification, I use excess deaths instead of excess ratios. This has no impact on the main message. But it is far less confusing to a reader that is not from this field.

[118] NASA Global climate change. Global Temperature. https://climate.nasa.gov/vital-signs/global-temperature/

[119] IPCC Special Report: Global warming of 1.5° C. https://www.ipcc.ch/sr15/

[120] IPCC Sixth Assessment Report (AR6). https://www.ipcc.ch/assessment-report/ar6/

[121] IPCC: AR6. Climate Change 2021: The Physical Science Basis. https://www.ipcc.ch/report/ar6/wg1/downloads/report/IPCC_AR6_WGI_SPM_final.pdf

[122] The IPCC report provides data relative to 1850-1900 (no warming). I share data relative to the present state (i.e., about 1°C warming). The extreme temperature events that occurred once every 10 years in absence of a warming occur 2.8 times every 10 years presently. The frequency will further increase by a factor of 3.4 for 4°C warming. So, extreme temperature events will occur 9.4 times every 10 years at a warming of 4°C. IPCC mentions that these are likely estimates.

[123] The IPCC report provides data relative to 1850-1900. I share data relative to the present state. The heavy precipitation events that occurred once in every 10 years in absence of a warming occur 1.3 times every 10 years presently (about 1°C warming). The frequency will further increase by a factor of 2.1 for 4°C warming. So, extreme temperature events will occur 2.7 times every 10 years at a warming of 4°C.

[124] The IPCC report provides data relative to 1850-1900. I share data relative to the present state. The drought events that occurred once in every 10 years in absence of a warming occur 1.7 times every 10 years presently (about 1°C warming). The frequency will further increase by a factor of 2.4 for 4°C warming. So, extreme temperature events will occur 4.1 times every 10 years at a warming of 4°C.

[125] UN Environment Programme. Emissions gap report 2023. https://www.unep.org/resources/emissions-gap-report-2023

[126] UN Environment Programme. Emissions gap report 2023. Nations must go further than current Paris pledges or face warming of 2.5°C to 2.9°C https://www.unep.org/news-and-stories/press-release/nations-must-go-further-current-paris-pledges-or-face-global-warming#:~:text=Nairobi%2C%2020%20November%202023%20–%20As,pre%2Dindustrial%20levels%20this%20century%2C

[127] IEA Report. World energy outlook 2023. https://www.iea.org/reports/world-energy-outlook-2023/executive-summary Stated policies scenario considers existing policies and those that are under development.

[128] World Bank data. GDP growth (annual %) World. https://data.worldbank.org/indicator/NY.GDP.MKTP.KD.ZG?locations=1W

[129] International Monetary Fund. World Economic Outlook, October 2019. https://www.imf.org/en/Publications/WEO/Issues/2019/10/01/world-economic-outlook-october-2019

[130] According to an IMF estimate, without COVID the GDP would have risen by 3%. Instead, the global GDP decreased by 3% because of the disease.

[131] IPCC Fifth Assessment Report-AR5 (2014). Key economic sectors and services. https://www.ipcc.ch/site/assets/uploads/2018/02/WGIIAR5-Chap10_FINAL.pdf See Summary (10.9.5).

[132] IPCC Sixth Assessment Report-AR6. Climate Change 2022. Impacts, adaptation, and vulnerability. Technical Summary. TS.C.10.2 (pg. 66-67). https://www.ipcc.ch/report/ar6/wg2/downloads/report/IPCC_AR6_WGII_SummaryVolume.pdf

[133] IPCC Fourth Assessment Report (2007). 20.6.1: History and present state of aggregate impact estimates. https://archive.ipcc.ch/publications_and_data/ar4/wg2/en/ch20s20-6-1.html

[134] IPCC Sixth Assessment Report-AR6. Climate Change 2022. Impacts, adaptation, and vulnerability. Technical Summary. TS.C.10.2 (pg. 66-67).

https://www.ipcc.ch/report/ar6/wg2/downloads/report/IPCC_AR6_WGII_SummaryVolume.pdf

[135] IPCC Fifth Assessment Report-AR5 (2014). Key economic sectors and services. https://www.ipcc.ch/site/assets/uploads/2018/02/WGIIAR5-Chap10_FINAL.pdf See Summary (10.9.5).

[136] IPCC Special Report: Global warming of 1.5°C. Chapter 3. Page 256.

[137] UN Environment Programme. Emissions gap report 2023. https://www.unep.org/resources/emissions-gap-report-2023

[138] UN Environment Programme. Emissions gap report 2023. Nations must go further than current Paris pledges or face warming of 2.5°C to 2.9°C https://www.unep.org/news-and-stories/press-release/nations-must-go-further-current-paris-pledges-or-face-global-warming#:~:text=Nairobi%2C%2020%20November%202023%20–%20As,pre%2Dindustrial%20levels%20this%20century%2C

[139] IEA Report. World energy outlook 2023. https://www.iea.org/reports/world-energy-outlook-2023/executive-summary Stated policies scenario considers existing policies and those that are under development.

[140] NASA Earth Observatory. Netherlands Dikes. https://earthobservatory.nasa.gov/images/5854/netherlands-dikes

[141] World bank Blogs (July 2023). https://blogs.worldbank.org/water/nature-based-solutions-netherlands-inspiration-improve-water-security

[142] IPCC: AR6. Climate Change 2021: The Physical Science Basis. Technical Summary. https://www.ipcc.ch/report/ar6/wg1/downloads/report/IPCC_AR6_WGI_TS.pdf

[143] My World: The United Nations Global Survey for a better world. We the peoples. https://www.un.org/youthenvoy/wp-content/uploads/2014/10/wethepeoples-7million.pdf

[144] My World: The United Nations Global Survey for a better world. Have your Say. https://vote.myworld2015.org

[145] The future we want. The United Nations we need. http://report.un75.online/files/report/un75-report-september-en.pdf

[146] United Nations Development Programme Report (2021). People's climate vote. Results. https://www.undp.org/publications/peoples-climate-vote

[147] People from fifty countries too part in the survey.

[148] According to the report, an average of 64% believed that climate change was an emergency. Of this subset, 59% agreed that everything necessary must be done urgently. So, the percent of total people that agreed that everything necessary must be done urgently equals 64*59% = 38%.

[149] Ipsos Survey (April 2023). Earth day 2023. Public opinion on climate change. https://www.ipsos.com/sites/default/files/ct/news/documents/2023-04/Ipsos%20Global%20Advisor%20-%20Earth%20Day%202023%20-%20Full%20Report%20-%20WEB.pdf

[150] Ipsos Survey (March 2023). What worries the world. https://www.ipsos.com/sites/default/files/ct/news/documents/2023-03/Global%20Report%20-%20What%20Worries%20the%20World%20Mar%2023.pdf

[151] UN Sustainable Development Goals. Website accessed Nov. 11, 2023. https://www.un.org/sustainabledevelopment/

[152] UN Sustainable Development Goals. Goal 6: Clean water and sanitation https://www.un.org/sustainabledevelopment/water-and-sanitation/

[153] UN Sustainable Development Goals. Goal 2: Zero Hunger. https://www.un.org/sustainabledevelopment/hunger/

[154] UN Sustainable Development Goals. Goal 4: Quality education. https://www.un.org/sustainabledevelopment/education/

[155] UN Sustainable Development Goals. Goal 7: Affordable and clean energy. https://www.un.org/sustainabledevelopment/energy/

[156] UN Sustainable Development Goals. Goal 3: Good health and well-being. https://www.un.org/sustainabledevelopment/health/

[157] UN IGME Report (2022). Levels and trends in child mortality. https://data.unicef.org/resources/levels-and-trends-in-child-mortality/

[158] World Bank data. https://pip.worldbank.org/home Data is in terms of 2017 PPP$. Meaning, it accounts for the difference in

the cost of living across countries. I input $10 for the poverty line to get the data. Website accessed on Nov. 11, 2023.

[159] Pew Research center. February 2023. Economy remains the publics top policy priority. https://www.pewresearch.org/politics/2023/02/06/economy-remains-the-publics-top-policy-priority-covid-19-concerns-decline-again/

[160] 2023 Associated Press-NORC/EPIC energy survey. Public opinion on energy and climate. https://epic.uchicago.edu/wp-content/uploads/2023/04/EPIC-Energy-Policy-Survey-2023_Topline.pdf

[161] Specifically, 16% strongly supported a 1$ additional cost per month while 22% somewhat supported it.

[162] R. Berner, The long-term carbon cycle, fossil fuels and atmospheric composition. Nature, 1, 5 (1981). https://www.nature.com/articles/nature02131

[163] U.S. EIA: Natural gas explained. https://www.eia.gov/energyexplained/natural-gas/

[164] Energy Institute. 2023 statistical review of world energy. https://www.energyinst.org/statistical-review

[165] Words rated. Number of academic papers published every year. https://wordsrated.com/number-of-academic-papers-published-per-year/#:~:text=As%20of%202022%2C%20over%205.14,5.03%20million%20papers%20were%20published. Over 64 million papers have been published since 1996 alone.

[166] Svante Arrhenius. Philosophical Magazine Series 5, 41, 251 (1896). https://www.tandfonline.com/doi/abs/10.1080/14786449608620846?src=recsys

[167] Guy Callendar. Quarterly Journal of the Royal Meteorological Society, 64, 223 (1938). https://www.rmets.org/sites/default/files/qjcallender38.pdf

[168] J. R. Fleming, Eos, Transactions American Geophysical Union 79, 405 (1998). https://agupubs.onlinelibrary.wiley.com/doi/abs/10.1029/98EO00310

[169] E. Hawkins, P.D. Jones, Quarterly Journal of the Royal Meteorological Society, 139, 1961 (2013). https://www.researchgate.net/publication/251231528_On_increasing_global_temperatures_75_years_after_Callendar https://centaur.reading.ac.uk/32981/1/hawkins_jones_2013.pdf

[170] Charles Keeling. Tellus, 12, 200 (1960). https://onlinelibrary.wiley.com/doi/abs/10.1111/j.2153-3490.1960.tb01300.x

[171] IPCC Fourth Assessment Report (AR4), 2007. Historical overview of climate change science. https://www.ipcc.ch/report/ar4/wg1/historical-overview-of-climate-change-science/

[172] Science, vol. 173, pg. 138-141 (1971). https://www.science.org/doi/10.1126/science.173.3992.138

[173] Bulletin of the American Meteorological society. Vol. 89, pg. 1325. https://journals.ametsoc.org/view/journals/bams/89/9/2008bams2370_1.xml

[174] NASA Global climate change. Global Temperature. https://climate.nasa.gov/vital-signs/global-temperature/

[175] NASA Global climate change. Global Temperature. https://climate.nasa.gov/vital-signs/global-temperature/

[176] Bulletin of the American Meteorological society. Vol. 89, pg. 1325. https://journals.ametsoc.org/view/journals/bams/89/9/2008bams2370_1.xml

[177] U.S. National Academy of Science Report: Understanding Climate Change (1975). https://archive.org/details/understandingcli00unit/mode/2up

[178] IPCC Fourth Assessment Report (AR4), 2007. Historical overview of climate change science. https://www.ipcc.ch/report/ar4/wg1/historical-overview-of-climate-change-science/

[179] IPCC history. https://www.ipcc.ch/about/history/

[180] Climate Change: The 1990 and 1992 IPCC Assessments. https://www.ipcc.ch/site/assets/uploads/2018/05/ipcc_90_92_assessments_far_overview.pdf

[181] NASA Global climate change. Global Temperature. https://climate.nasa.gov/vital-signs/global-temperature/

[182] IPCC Sixth Assessment Report-AR6 (2021). Frequently Asked Questions. FAQ 3.2 https://www.ipcc.ch/report/ar6/wg1/downloads/faqs/IPCC_AR6_WGI_FAQ_Chapter_03.pdf

[183] IPCC reports. https://www.ipcc.ch/reports/

[184] U.S. EIA: Natural Gas and the Environment. https://www.eia.gov/energyexplained/natural-gas/natural-gas-and-the-environment.php

[185] IEA: Role of gas in today's energy transition. https://www.iea.org/reports/the-role-of-gas-in-todays-energy-transitions#key-findings

[186] U.S. National Energy technology Laboratory: Life cycle greenhouse gas emissions. Natural gas and power production. https://www.eia.gov/conference/2015/pdf/presentations/skone.pdf

[187] Annex III: Technology specific cost and performance parameters. Climate Change 2014: Mitigation of Climate Change. Contribution of Working Group III to the Fifth Assessment Report of the Intergovernmental Panel on Climate Change. https://www.ipcc.ch/site/assets/uploads/2018/02/ipcc_wg3_ar5_annex-iii.pdf#page=7

[188] IEA: Comparative life cycle GHG emissions from a midsize BEV and ICE vehicle (2021). https://www.iea.org/data-and-statistics/charts/comparative-life-cycle-greenhouse-gas-emissions-of-a-mid-size-bev-and-ice-vehicle This estimate is based on an average carbon intensity of the global electrical grid. Carbon intensity is the amount of greenhouse gases produced per unit of electricity generated.

[189] The 50% reduction is too optimistic based on recent studies. Prior studies used the average carbon intensity of the grid. Of more relevance is the carbon intensity of the marginal electricity being used by electric cars. The marginal electricity is typically provided by coal power or natural gas. So, the carbon intensity of the marginal electricity is higher. This translates to a much lower benefit from electric cars. https://www.pnas.org/doi/abs/10.1073/pnas.2116632119

[190] U.S. EIA: Natural Gas and the Environment. https://www.eia.gov/energyexplained/natural-gas/natural-gas-and-the-environment.php

[191] U.S. EIA: Cost and performance characteristics of new generating technologies (2020). https://www.eia.gov/outlooks/aeo/assumptions/pdf/table_8.2.pdf

[192] IEA Report: Projected costs of generating electricity 2020. https://www.iea.org/reports/projected-costs-of-generating-electricity-2020

[193] T. V. Choudhary. Critical Comparison of Low Carbon Technologies (October 2020). https://www.amazon.com/dp/B08LP8TRLP

[194] U.S. Energy Information Administration: Levelized cost of new generation resources in the annual energy outlook 2022. https://www.eia.gov/outlooks/aeo/pdf/electricity_generation.pdf

[195] International Renewable Energy Agency (IRENA). Flexibility in conventional power plants. https://www.irena.org/-/media/Files/IRENA/Agency/Publication/2019/Sep/IRENA_Flexibility_in_CPPs_2019.pdf?la=en&hash=AF60106EA083E492638D8FA9ADF7FD099259F5A1

[196] Annex III: Technology specific cost and performance parameters. Climate Change 2014: Mitigation of Climate Change. Contribution of Working Group III to the Fifth Assessment Report of the Intergovernmental Panel on Climate Change. https://www.ipcc.ch/site/assets/uploads/2018/02/ipcc_wg3_ar5_annex-iii.pdf#page=7

[197] U.S. National Renewable Energy Laboratory 2021 Update: Lifecycle greenhouse gas emissions from electricity generation. September 2021. https://www.nrel.gov/docs/fy21osti/80580.pdf

[198] United Nations Economic Commission for Europe (March 2022). Lifecycle assessments of electricity generation options. https://unece.org/sed/documents/2021/10/reports/life-cycle-assessment-electricity-generation-options

[199] IEA: Methane emissions from oil and gas. https://www.iea.org/reports/methane-emissions-from-oil-and-gas-operations

[200] U.S. EPA. Dec. 16 Report: Hydraulic fracturing for oil and gas: Impacts from the hydraulic fracturing water cycle on drinking

water resources in the United States. https://cfpub.epa.gov/ncea/hfstudy/recordisplay.cfm?deid=332990

[201] U.S. EPA. Unconventional oil and natural gas development. https://www.epa.gov/uog Website accessed Nov. 12, 2023.

[202] U.S. Geological Survey: Who we are? https://www.usgs.gov/about/about-us/who-we-are

[203] U.S. EIA: The distribution of U.S. oil and natural gas wells according to production rates. https://www.eia.gov/petroleum/wells/pdf/full_report.pdf

[204] U.S. Geological Survey: Hydraulic fracturing. https://www.usgs.gov/mission-areas/water-resources/science/hydraulic-fracturing?qt-science_center_objects=0#overview

[205] U.S. Geological Survey. https://www.usgs.gov/search?keywords=hydraulic+fracturing

[206] U.S. Geological Survey, (May 2017). https://www.usgs.gov/news/national-news-release/unconventional-oil-and-gas-production-not-currently-affecting-drinking

[207] U.S. Geological Survey: Water resources (December 2020). https://www.usgs.gov/news/bakken-shale-unconventional-oil-and-gas-production-has-not-caused-widespread-hydrocarbon

[208] U.S. Geological Survey: Water resources (July 2019). https://www.usgs.gov/news/marcellus-shale-natural-gas-production-not-currently-causing-widespread-hydrocarbon

[209] U.S. Geological Survey, (May 2017). Does fracking cause earthquakes? https://www.usgs.gov/faqs/does-fracking-cause-earthquakes#

[210] Health Effects Institute (HEI) is a non-profit organization chartered in 1980 as an independent research organization to provide impartial science on the health effects of air pollution. It is funded by U.S. EPA and worldwide motor vehicle industry. https://www.healtheffects.org/about

[211] World Health organization (WHO): Ambient outdoor Air pollution. https://www.who.int/news-room/fact-sheets/detail/ambient-(outdoor)-air-quality-and-health

[212] Health Effects Institute (HEI): State of global air 2020. https://www.stateofglobalair.org HEI is funded by U.S. EPA and worldwide motor vehicle industry.

[213] Links to references. https://www.ncbi.nlm.nih.gov/pmc/articles/PMC2568866/ ; https://www.frontiersin.org/articles/10.3389/fenvs.2022.996038/full#B74 ; https://www.jacionline.org/article/S0091-6749(19)30582-2/fulltext ; https://academic.oup.com/ije/article/32/5/847/665734 ; https://ourworldindata.org/indoor-air-pollution ; https://www.pnas.org/doi/10.1073/pnas.1019183108 ; https://www.sciencedirect.com/science/article/pii/S2212609014000521 ; https://oem.bmj.com/content/66/11/777

[214] Nature Communications, 12,3594, 2021. https://www.nature.com/articles/s41467-021-23853-y#Abs1 The study is a collaboration between authors from Washington University, Dalhousie University, Spadaro Environmental Research Consultants, IHME, University of Washington, Pacific Northwest National Laboratory, Department of Atmospheric Sciences, University of Washington, University at Albany, Peking University and University of British Columbia.

[215] Health Effects Institute (HEI): State of global air. The global impact of fossil fuels (Nov. 2023). https://www.stateofglobalair.org/resources/storymap/air-pollution-and-health

[216] The data was for the year 2019. But it is a good indicator for current deaths per year because the air pollution from fossil fuels does not change much from year to year. Health Effects Institute (HEI): State of global air. Global burden of diseases from major air pollution sources. Research report 210. https://www.healtheffects.org/publication/global-burden-disease-major-air-pollution-sources-gbd-maps-global-approach

[217] The LANCET Countdown. November 14, 2023. The 2023 report of the LANCET countdown on health and climate change: the imperative for a health centered response in a world facing irreversible harms. https://www.thelancet.com/journals/lancet/article/PIIS0140-6736(23)01859-7/fulltext

[218] Environmental Research. Vol. 125, 110754 (2021). https://www.sciencedirect.com/science/article/abs/pii/S0013935121000487?via%3Dihub)

[219] The LANCET Countdown. November 14, 2023. The 2023 report of the LANCET countdown on health and climate change: the imperative for a health centered response in a world facing irreversible harms. https://www.thelancet.com/journals/lancet/article/PIIS0140-6736(23)01859-7/fulltext DOI:

[220] I have discussed the sources earlier in this section.

[221] Nature Communications, 12,3594, 2021. https://www.nature.com/articles/s41467-021-23853-y#Abs1 The study is a collaboration between authors from Washington University, Dalhousie University, Spadaro Environmental Research Consultants, IHME, University of Washington, Pacific Northwest National Laboratory, Department of Atmospheric Sciences, University of Washington, University at Albany, Peking University and University of British Columbia.

[222] The LANCET Countdown. November 14, 2023. The 2023 report of the LANCET countdown on health and climate change: the imperative for a health centered response in a world facing irreversible harms. https://www.thelancet.com/journals/lancet/article/PIIS0140-6736(23)01859-7/fulltext

[223] Nature Communications, 12,3594, 2021. https://www.nature.com/articles/s41467-021-23853-y#Abs1 The study is a collaboration between authors from Washington University, Dalhousie University, Spadaro Environmental Research Consultants, IHME, University of Washington, Pacific Northwest National Laboratory, Department of Atmospheric Sciences, University of Washington, University at Albany, Peking University and University of British Columbia.

[224] The LANCET Countdown. November 14, 2023. The 2023 report of the LANCET countdown on health and climate change: the imperative for a health centered response in a world facing irreversible harms. https://www.thelancet.com/journals/lancet/article/PIIS0140-6736(23)01859-7/fulltext

[225] World Health organization (WHO): Tobacco, Key facts. https://www.who.int/news-room/fact-sheets/detail/tobacco

[226] World Health organization (WHO): Ambient outdoor Air pollution. https://www.who.int/news-room/fact-sheets/detail/ambient-(outdoor)-air-quality-and-health

[227] Health Effects Institute (HEI): State of global air 2020. https://www.stateofglobalair.org HEI is funded by U.S. EPA and worldwide motor vehicle industry.

[228] World Health organization (WHO): Obesity. https://www.who.int/news-room/facts-in-pictures/detail/6-facts-on-obesity

[229] Centre for Disease Control and Prevention (CDC). Road traffic injuries and death: A global problem. https://www.cdc.gov/injury/features/global-road-safety/index.html

[230] World Trade organization. Defining subsidies. https://www.wto.org/english/res_e/booksp_e/anrep_e/wtr06-2b_e.pdf

[231] Cambridge Dictionary. https://dictionary.cambridge.org/us/dictionary/english/subsidy

[232] This is the average per year subsidy received by fossil fuels over the past decade.

[233] UNDP: Fossil fuel subsidy reforms, lessons, and opportunities (2021). https://www.undp.org/publications/fossil-fuel-subsidy-reform-lessons-and-opportunities

[234] IEA: Energy subsidies. https://www.iea.org/topics/energy-subsidies

[235] A large fraction of the total subsidies are energy not specifically fossil fuel subsidies. Fossil fuels receive most of the energy subsidies because they have a massive share in the global energy mix. According to UNDP, most of these subsidies end up with the high-income households, thus defeating the purpose. Reforms to the subsidies are, therefore, important. UNDP: Fossil fuel subsidy reforms, lessons, and opportunities (2021). https://www.undp.org/publications/fossil-fuel-subsidy-reform-lessons-and-opportunities A large fraction of the total subsidies are energy subsidies. Fossil fuels receive the majority of the

energy subsidies because they have a massive share in the global energy mix.

[236] International Monetary Fund report. IMF fossil fuel subsidies data. 2023 Update. https://www.imf.org/en/Publications/WP/Issues/2023/08/22/IMF-Fossil-Fuel-Subsidies-Data-2023-Update-537281

[237] IMF has released several such reports. Typically, there is an update each year. The previous reference is for the latest update.

[238] There is no dispute about the importance of these side-effects. They are very important. But it does not make sense to include the cost of the side-effects as subsidies. I discuss why in the text below.

[239] The report uses the term “implicit subsidies” for side-effects. It uses the term “explicit subsidies” for true subsidies (the common definition).

[240] The global society consumes a massive amount of energy. Enormous resources will be required even if it is produced from non-fossil sources. Large resources equal to a large impact on the environment. I discuss details in a later chapter.

[241] IEA Updated 2023 report: Net zero by 2050. https://www.iea.org/reports/net-zero-roadmap-a-global-pathway-to-keep-the-15-0c-goal-in-reach

[242] IEA. Reaching net zero emissions requires faster innovation, but we have already come a long way. https://www.iea.org/commentaries/reaching-net-zero-emissions-demands-faster-innovation-but-weve-already-come-a-long-way

[243] The global society consumes a massive amount of energy. Enormous resources will be required even if it is produced from non-fossil sources. Large resources equal to a large impact on the environment. I discuss details in a later chapter.

[244] My discussion is not an argument against the proposed reforms in the report. The problem is that the proposal in the report is based on a very poor analysis. This destroys the credibility of the proposal.

[245] U.S. EPA: Benefits and cost of the clean air act 1990-2020. https://www.epa.gov/clean-air-act-overview/benefits-and-costs-clean-air-act-1990-2020-second-prospective-study

[246] If the society chooses to, it can lower the deaths and other impacts from pollution very cost-effectively using pollution control technologies. Most of the cost of this side-effect is in the hands of the global society. This report assumes otherwise.

[247] I will discuss an example of air pollution control from coal power plants in the United States. PM2.5 particulate matter pollution was decreased by 93% per unit of electricity generated by coal power from 1999 to 2020. This decreased the pollution related deaths by 97% over that period.
https://www.science.org/doi/10.1126/science.adf4915
https://www.eia.gov/energyexplained/electricity/electricity-in-the-us-generation-capacity-and-sales.php

[248] The details about the credible studies are discussed in the previous section.

[249] The LANCET Countdown. November 14, 2023. The 2023 report of the LANCET countdown on health and climate change: the imperative for a health centered response in a world facing irreversible harms.
https://www.thelancet.com/journals/lancet/article/PIIS0140-6736(23)01859-7/fulltext

[250] Nature Communications, 12,3594, 2021.
https://www.nature.com/articles/s41467-021-23853-y#Abs1

[251] Health Effects Institute (HEI). Global burden of disease from major air pollution sources. A global approach.
https://www.ncbi.nlm.nih.gov/pmc/articles/PMC9501767/
https://pubmed.ncbi.nlm.nih.gov/36148817/

[252] The credible studies (prior two references) indicate the number of deaths to be around 1 million per year. One of the authors of these papers, Michael Brauer, is also a major contributor to the key HEI report on this topic.

[253] The assumed cost per death is adjusted based on the economic state of a country. It is much lower in developing countries.

[254] At the bare minimum, the report should emphasize that the road congestion and accidents are related to personal transportation and do not depend on the vehicle technology. This should be highlighted several times in the report.

[255] IPCC Special Report: Global warming of 1.5 °C. Chapter 2.
https://www.ipcc.ch/sr15/chapter/chapter-2/

[256] NOAA. Ocean-atmosphere CO_2 exchange. https://sos.noaa.gov/datasets/ocean-atmosphere-CO2-exchange/

[257] U.S. Department of Energy (DOE). DOE explains isotopes. https://www.energy.gov/science/doe-explainsisotopes

[258] NOAA Global Monitoring Laboratory. The Basics: Isotopic Fingerprints. https://gml.noaa.gov/outreach/isotopes/

[259] IPCC Climate change 2013 (AR5). The Physical Basis. Chapter 6. Section 6.3.2.3 https://www.ipcc.ch/site/assets/uploads/2018/02/WG1AR5_Chapter06_FINAL.pdf

[260] United Nations. Causes and effects of climate change. https://www.un.org/en/climatechange/science/causes-effects-climate-change

[261] IPCC Reports. https://www.ipcc.ch/reports/

[262] PBL Netherlands Environmental Assessment Agency (2022). Trends in global CO_2 and total greenhouse gas emissions. https://www.pbl.nl/en/trends-in-global-CO2-emissions

[263] IEA Technology Report. Global energy review: CO_2 emissions in 2021. Power sector: 14.6 billion tons. Energy sector: 36.3 billion tons. https://www.iea.org/reports/global-energy-review-CO2-emissions-in-2021-2

[264] Climate Watch: Historical GHG emissions. https://www.climatewatchdata.org/ghg-emissions?breakBy=regions&end_year=2018&gases=ch4®ions=WORLD%2CWORLD§ors=electricity-heat&start_year=1990

[265] Power sector contributes to 30% the global greenhouse gas emissions.

[266] U.S. Energy Information Administration: Levelized cost of new generation resources in the annual energy outlook 2021. https://www.eia.gov/outlooks/aeo/pdf/electricity_generation.pdf

[267] U.S. Energy Information Administration: Levelized cost of new generation resources in the annual energy outlook 2022. https://www.eia.gov/outlooks/aeo/pdf/electricity_generation.pdf

[268] National Renewable Energy Laboratory (2015): Overgeneration from solar energy in California. https://www.nrel.gov/docs/fy16osti/65023.pdf

[269] National Renewable Energy Laboratory: Ten years of analyzing the duck chart. https://www.nrel.gov/news/program/2018/10-years-duck-curve.html

[270] Current Result: Weather and Science Facts. Sunniest places in the world. https://www.currentresults.com/Weather-Extremes/sunniest-places-countries-world.php

[271] World Meteorological Department (WMO). Measuring sunlight. For example, the sunniest year in Germany had an average sunshine of five and a half hours per day https://public.wmo.int/en/measuring-sunlight

[272] OECD and NEA report (2012): Nuclear Energy and Renewables. System effects in low carbon low carbon electricity systems. https://www.oecd.org/publications/nuclear-energy-and-renewables-9789264188617-en.htm

[273] IEA: Projected costs of generating electricity 2020. https://www.iea.org/reports/projected-costs-of-generating-electricity-2020

[274] OECD NEA report (2019): System costs with high share of nuclear energy and renewables. https://www.oecd-nea.org/jcms/pl_15000/the-costs-of-decarbonisation-system-costs-with-high-shares-of-nuclear-and-renewables?details=true

[275] Energy and Environmental Science (2022). Low-cost solutions to global warming, air pollution and energy security for 145 countries. https://pubs.rsc.org/en/content/articlelanding/2022/ee/d2ee00722c

[276] Joule (2022). Empirically grounded technology forecasts and the energy transition. https://doi.org/10.1016/j.joule.2022.08.009

[277] OECD NEA report (2019): System costs with high share of nuclear energy and renewables. https://www.oecd-nea.org/jcms/pl_15000/the-costs-of-decarbonisation-system-costs-with-high-shares-of-nuclear-and-renewables?details=true

[278] Solar energy is the direct source for solar power. Solar energy unevenly heats earth, which leads to wind generation. Fossil fuels have formed from the remains of dead organisms, i.e., stored solar energy.

[279] This is a well-established principle. For example, the cost of building a home at a given location depends on the resource

intensity. A home that requires more labor, materials and other resources will be more expensive.

[280] Certain technologies require a continuous supply of fossil fuels. Others require large amount of land and materials. Also, the technologies require different types of materials for the construction of power plants. The resources are not directly comparable. They cannot be directly added.

[281] Energy policy, 123, 83, 2018. https://www.researchgate.net/publication/327239302_The_spatial_extent_of_renewable_and_non-renewable_power_generation_A_review_and_meta-analysis_of_power_densities_and_their_application_in_the_US

[282] International journal of Green Energy, 5,438, 2008. https://www.researchgate.net/publication/233231163_A_Comparison_of_Energy_Densities_of_Prevalent_Energy_Sources_in_Units_of_Joules_Per_Cubic_Meter

[283] U.S. EIA: Solar explained. https://www.eia.gov/energyexplained/solar/

[284] U.S. EIA: Natural gas explained. https://www.eia.gov/energyexplained/natural-gas/

[285] Energy policy, 123, 83, 2018. https://www.researchgate.net/publication/327239302_The_spatial_extent_of_renewable_and_non-renewable_power_generation_A_review_and_meta-analysis_of_power_densities_and_their_application_in_the_US

[286] International journal of Green Energy, 5,438, 2008. https://www.researchgate.net/publication/233231163_A_Comparison_of_Energy_Densities_of_Prevalent_Energy_Sources_in_Units_of_Joules_Per_Cubic_Meter

[287] Energy policy, 123, 83, 2018. https://www.researchgate.net/publication/327239302_The_spatial_extent_of_renewable_and_non-renewable_power_generation_A_review_and_meta-analysis_of_power_densities_and_their_application_in_the_US

[288] International journal of Green Energy, 5,438, 2008. https://www.researchgate.net/publication/233231163_A_Comparison_of_Energy_Densities_of_Prevalent_Energy_Sources_in_Units_of_Joules_Per_Cubic_Meter

[289] Long duration storage is not required currently because the electrical grids are dominated by dispatchable technologies. Solar and wind power only provide a small fraction of electricity. This will change in a net zero world. All the proposals suggest that solar and wind power will dominate the electrical grids in a net zero world.

[290] Currently, the project cost of solar and wind power plant with just 12 hours of battery storage is five times more than the cost of a natural gas power plant to provide the same amount of annual electricity. The practical energy storage needs for most regions will be 24+ hours to many days. Thus, the project costs with current technology will be extremely high. Data for costs is from the following references. https://www.eia.gov/outlooks/aeo/assumptions/pdf/elec_cost_perf.pdf https://www.nrel.gov/docs/fy23osti/85332.pdf Green hydrogen is even more costly.

[291] T. V. Choudhary. Critical Comparison of Low Carbon Technologies (October 2020). https://www.amazon.com/dp/B08LP8TRLP

[292] Upfront cost of concentrated solar with thermal for only 12 hours of storage is also about five times more than natural gas power plants. This is the low-end cost. Energy storage that is substantially longer than 12 hours would be required for providing 24x7 electricity. https://www.iea.org/reports/projected-costs-of-generating-electricity-2020

[293] U.S. Department of Energy. The pathway to long duration energy storage commercial lift off. https://liftoff.energy.gov/long-duration-energy-storage/

[294] Sustainability. A review of the current and past state of EROI data. Volume 3. Page 1796. Year 2011. https://www.mdpi.com/2071-1050/3/10/1796

[295] Sustainability. EROI of major energy carriers. Review and harmonization. Volume 14. Page 7098. Year 2022. https://www.mdpi.com/2071-1050/14/12/7098?trk=public_post_comment-text

[296] Certain technologies require a continuous supply of fossil fuels. Others require large amount of land and materials. Also, the technologies require different types of materials for the

construction of power plants. The resources are not directly comparable. They cannot be directly added.

[297] Brainy quote. Albert Einstein quotes. https://www.brainyquote.com/quotes/albert_einstein_106912

[298] International Energy Agency: Projected costs of generating electricity, 2020 Edition. https://www.iea.org/reports/projected-costs-of-generating-electricity-2020

[299] Lazard. Lazard's LCOE analysis. Version 15. https://www.lazard.com/perspective/levelized-cost-of-energy-levelized-cost-of-storage-and-levelized-cost-of-hydrogen/

[300] The Energy & Resources Institute (2019): Exploring electricity supply mix scenarios to 2030. https://www.teriin.org/sites/default/files/2019-02/Exploring%20Electricity%20Supply-Mix%20Scenarios%20to%202030.pdf

[301] National Renewable Energy Laboratory (October 2021). Fall 2021 Solar Industry Update. https://www.nrel.gov/docs/fy22osti/81325.pdf

[302] National Renewable Energy Laboratory (September 2022): U.S. Solar photovoltaic system cost and energy storage benchmark: Q1 2022. https://www.nrel.gov/docs/fy22osti/83586.pdf

[303] Nature Energy. It is time for rooftop solar to compete with other renewables. Vol. 7. Page 298. Year 2022.

[304] Energypost.eu, October 2020.Why promote rooftop solar when the grid is so cheaper? https://energypost.eu/why-promote-rooftop-solar-when-the-grid-is-so-much-cheaper/

[305] The Energy Institute Blog, UC Berkeley, and cited references. Does rooftop solar help the distribution system? https://energyathaas.wordpress.com/2018/06/25/does-rooftop-solar-help-the-distribution-system/

[306] National Renewable Energy Laboratory (September 2022): U.S. Solar photovoltaic system cost and energy storage benchmark: Q1 2022. https://www.nrel.gov/docs/fy22osti/83586.pdf

[307] Britannica. Economy of scale. https://www.britannica.com/topic/economy-of-scale

[308] Lazard. Lazard's LCOE analysis. Version 15. https://www.lazard.com/perspective/levelized-cost-of-energy-levelized-cost-of-storage-and-levelized-cost-of-hydrogen/

[309] DSIRE. Database of state incentives for renewables and efficiency. https://www.dsireusa.org

[310] U.S. EIA: Electricity data. https://www.eia.gov/international/data/world/electricity/electricity-generation

[311] International Energy Agency (September 2022). Renewable electricity. https://www.iea.org/reports/renewable-electricity

[312] Energy Institute. 2023 Statistical review of world energy. https://www.energyinst.org/statistical-review

[313] U.S. EIA: Electricity data. https://www.eia.gov/international/data/world/electricity/electricity-generation

[314] International Energy Agency (September 2022). Electricity Sector. https://www.iea.org/reports/electricity-sector

[315] BP Statistical review of World energy 2022. https://www.bp.com/content/dam/bp/business-sites/en/global/corporate/pdfs/energy-economics/statistical-review/bp-stats-review-2022-full-report.pdf

[316] EMBER. World data (2023). https://ember-climate.org/countries-and-regions/regions/world/

[317] U.S. EIA: Electricity data. https://www.eia.gov/international/data/world/electricity/electricity-generation

[318] For reference, an average home in the United States uses about 30 kWh of electricity per day. U.S. EIA. https://www.eia.gov/tools/faqs/faq.php?id=97&t=3

[319] U.S. EIA: Electricity data. https://www.eia.gov/international/data/world/electricity/electricity-generation

[320] IEA. Wind Energy in Denmark. https://iea-wind.org/about-iea-wind-tcp/members/denmark/

[321] Danish Energy Agency. Integration of wind energy in power systems. A Danish experience (May 2017). https://ens.dk/sites/ens.dk/files/Globalcooperation/integration_of_wind_energy_in_power_systems.pdf

[322] Danish Energy Agency. Integration of wind energy in power systems. A Danish experience (May 2017).

https://ens.dk/sites/ens.dk/files/Globalcooperation/integration_of_wind_energy_in_power_systems.pdf

[323] Ember, 2021 Global Electricity data review. https://ember-climate.org/data/global-electricity/

[324] Alternatively, it would require very costly long duration energy storage. The high cost can be avoided in Denmark only because of the unique conditions.

[325] IEA report. Grid scale storage. https://www.iea.org/reports/grid-scale-storage#

[326] EIA today in energy. Hourly electric consumption varies throughout the day and across the seasons. https://www.eia.gov/todayinenergy/detail.php?id=42915

[327] U.S. EIA data. Hourly electric grid monitor. Data is from the United States for June 21, 2023. https://www.eia.gov/electricity/gridmonitor/dashboard/electric_overview/US48/US48

[328] U.S. EIA data. Hourly electric grid monitor. Data is from the United States for June 21, 2023. https://www.eia.gov/electricity/gridmonitor/dashboard/electric_overview/US48/US48

[329] U.S. EIA. Levelized cost of new generation resources in the annual energy outlook 2022. https://www.eia.gov/outlooks/aeo/pdf/electricity_generation.pdf

[330] U.S. EIA Annual energy outlook 2022. Footnote 3. https://www.eia.gov/outlooks/aeo/electricity_generation.php

[331] International Renewable Energy Agency. Flexibility in conventional power plants. https://www.irena.org/-/media/Files/IRENA/Agency/Publication/2019/Sep/IRENA_Flexibility_in_CPPs_2019.pdf?la=en&hash=AF60106EA083E492638D8FA9ADF7FD099259F5A1

[332] But there are limits to the flexibility. For example, a natural gas power plant cannot be operated below roughly 30%.

[333] IEA report. Grid scale storage. https://www.iea.org/reports/grid-scale-storage#

[334] IEA: Projected costs of generating electricity 2020. https://www.iea.org/reports/projected-costs-of-generating-electricity-2020

[335] OECD NEA report (2019). System costs with high share of nuclear energy and renewables. https://www.oecd-nea.org/jcms/pl_15000/the-costs-of-decarbonisation-system-costs-with-high-shares-of-nuclear-and-renewables?details=true

[336] EIA today in energy. June 21, 2023. As solar capacity grows, duck curves are getting deeper in California. https://www.eia.gov/todayinenergy/detail.php?id=56880

[337] National Renewable Energy Laboratory Report. Ten years of analyzing the duck chart. https://www.nrel.gov/news/program/2018/10-years-duck-curve.html

[338] To ramp down is to decrease production.

[339] IEA data. Hourly electric grid monitor. Data is from the United States for June 21, 2023. https://www.eia.gov/electricity/gridmonitor/dashboard/electric_overview/US48/US48

[340] California Energy commission. 2021 total system electric generation. https://www.energy.ca.gov/data-reports/energy-almanac/california-electricity-data/2021-total-system-electric-generation

[341] California ISO. Managing oversupply. http://www.caiso.com/informed/Pages/ManagingOversupply.aspx

[342] For example, power lines do not have enough capacity to deliver electricity.

[343] U.S. EIA. Today in energy. Solar and wind power curtailments are rising in California. https://www.eia.gov/todayinenergy/detail.php?id=60822#:~:text=As%20of%20September%2C%20CAISO%20has,first%20seven%20months%20of%202023.

[344] U.S. EIA data. Electricity. https://www.eia.gov/international/data/world/electricity/electricity-consumption

[345] World Bank data. Population of Chad. https://data.worldbank.org/indicator/SP.POP.TOTL?locations=TD

[346] California Energy commission. 2021 total system electric generation. https://www.energy.ca.gov/data-reports/energy-

almanac/california-electricity-data/2021-total-system-electric-generation

[347] California ISO. Our evolving grid. http://www.caiso.com/about/Pages/Blog/Posts/Our-Evolving-Grid.aspx

[348] California Energy commission. 2021 total system electric generation. https://www.energy.ca.gov/data-reports/energy-almanac/california-electricity-data/2021-total-system-electric-generation

[349] EIA today in energy. June 21, 2023. As solar capacity grows, duck curves are getting deeper in California. https://www.eia.gov/todayinenergy/detail.php?id=56880

[350] FERC report. November 16, 2021. https://www.ferc.gov/media/february-2021-cold-weather-outages-texas-and-south-central-united-states-ferc-nerc-and

[351] EIA today in energy. June 21, 2023. As solar capacity grows, duck curves are getting deeper in California. https://www.eia.gov/todayinenergy/detail.php?id=56880

[352] North American Electric Reliability Corporation. 2023 ERO reliabilities priority report. https://www.nerc.com/comm/RISC/Related%20Files%20DL/RISC_ERO_Priorities_Report_2023_Board_Approved_Aug_17_2023.pdf

[353] OECD NEA report: Nuclear Energy and Renewables. System effects in low carbon electricity systems.https://www.oecd.org/publications/nuclear-energy-and-renewables-9789264188617-en.htm

[354] OECD NEA report (2019): System costs with high share of nuclear energy and renewables. https://www.oecd-nea.org/jcms/pl_15000/the-costs-of-decarbonisation-system-costs-with-high-shares-of-nuclear-and-renewables?details=true

[355] The share of solar and wind power in California generation mix is 25%. https://www.energy.ca.gov/data-reports/energy-almanac/california-electricity-data/2021-total-system-electric-generation This is already causing major problems despite. Grids which are less flexible will have much larger problems for 25% share of solar and wind power. California is planning major

transmission line upgrades and storage projects to address this issue. But such solutions add substantial costs to the system.

[356] Any grid with high levels of solar and wind power will face the same problem as the California grid. It will produce far more electricity than it needs during some periods. It will produce far less electricity than it needs during other periods. So, all the grids with a high share of solar and wind power will be in a similar situation.

[357] Energies. A brief climatology of dunkelflaute events over and surrounding the North and Baltic Sea areas. Vol. 14, Page 6508, Year 2021. https://doi.org/10.3390/en14206508

[358] There is a large variation in the amount of electricity produced from solar and wind power over time because of weather conditions. Consider the performance of solar and wind power over the past year in the United States as an example. There were consecutive days (two or more) where the combined output from solar and wind was drastically reduced in most of its grids. This was also true for the whole of the United States. https://www.eia.gov/electricity/gridmonitor/dashboard/electric_overview/US48/US48

[359] IEA: United Net zero by 2050. https://www.iea.org/reports/net-zero-by-2050

[360] IRENA: World transitions outlook: 1.5°C pathway. https://irena.org/publications/2021/Jun/World-Energy-Transitions-Outlook

[361] The cost will be prohibitive to maintain a large fleet of dispatchable plants for such occasions.

[362] Electrification is a shift to technologies that use electricity. Examples are electric vehicles and heat pumps.

[363] IEA: United Net zero by 2050. https://www.iea.org/reports/net-zero-by-2050

[364] Long duration storage can extend from days to weeks.

[365] Consider the performance of solar and wind power over the past year in the United States as an example. There were consecutive days (two or more) where the combined output from solar and wind was drastically reduced in most of its grids. This was also true for the combined United States grid.

https://www.eia.gov/electricity/gridmonitor/dashboard/electric_overview/US48/US48

[366] Nature Communications. Vol. 12. Article # 6146. Year 2021. https://www.nature.com/articles/s41467-021-26355-z

[367] There will be numerous practical challenges for operating continental scale grids. An unprecedented level of co-operation will be required between countries. The costs will also be high because of the massive size of the grids and large energy storage requirements.

[368] Details are provided in an earlier section.

[369] There are many practical challenges for these options. For example, the vehicle to grid option requires a reliance on individual car owners to maintain grid reliability. It encourages private car ownership, which is ineffective and problematic for the environment. It assumes that individual owners will charge and discharge their car batteries in an optimal manner. Alternatively, it assumes that there will be a central control of how privately owned vehicles are charged and discharged. Overall, it is not practical and will soundly be rejected by the global majority.

[370] U.S. EIA: Electricity data. https://www.eia.gov/international/data/world/electricity/electricity-generation Average share of nuclear power from 1980 to 2021 was 14%.

[371] U.S. EIA: Nuclear power plants. https://www.eia.gov/energyexplained/nuclear/nuclear-power-plants.php

[372] U.S. EIA: Electricity data. https://www.eia.gov/international/data/world/electricity/electricity-generation

[373] The Lancet (2007). Electricity generation and health. Vol. 370; Pg. 979. https://pubmed.ncbi.nlm.nih.gov/17876910/

[374] Journal of cleaner production (2016). Balancing safety with sustainability. Vol. 112; Pg. 3952. https://www.sciencedirect.com/science/article/abs/pii/S0959652615009877

[375] Our World in Data. Nuclear Energy. https://ourworldindata.org/nuclear-energy

[376] Data summary from Our World in Data. Deaths related to different technologies per TWh of electricity produced. Wind: 0.04; Solar: 0.02; Nuclear: 0.03; Coal: > 20. https://ourworldindata.org/nuclear-energy

[377] Three examples are the Fukushima (2011), Chernobyl (1986) and Three Mile Island (1979) accidents.

[378] U.S. DOE-Office of nuclear energy: Advantages and Challenges of nuclear energy. https://www.energy.gov/ne/articles/advantages-and-challenges-nuclear-energy

[379] IEA Energy Technology Perspectives 2023. https://www.iea.org/reports/energy-technology-perspectives-2023#

[380] Oakridge National Laboratory Report. Environmental quality and U.S. power sector. Air quality, water quality, land use and environmental justice. https://www.energy.gov/sites/prod/files/2017/01/f34/Environment%20Baseline%20Vol.%202--Environmental%20Quality%20and%20the%20U.S.%20Power%20Sector--Air%20Quality%2C%20Water%20Quality%2C%20Land%20Use%2C%20and%20Environmental%20Justice.pdf

[381] Unlike fossil fuel power, solar and wind power do not require hydrocarbon fuel for generating electricity.

[382] U.S. National Renewable Energy Laboratory 2021 Update: Lifecycle greenhouse gas emissions from electricity generation. September 2021. https://www.nrel.gov/docs/fy21osti/80580.pdf

[383] United Nations Economic Commission for Europe (March 2022). Lifecycle assessments of electricity generation options. https://unece.org/sed/documents/2021/10/reports/life-cycle-assessment-electricity-generation-options

[384] U.S. National Energy Technology Laboratory: Life cycle greenhouse gas emissions. Natural gas and power production. https://www.eia.gov/conference/2015/pdf/presentations/skone.pdf

[385] Annex III: Technology specific cost and performance parameters. Climate Change 2014: Mitigation of Climate Change. Contribution of Working Group III to the Fifth Assessment

Report of the Intergovernmental Panel on Climate Change. https://www.ipcc.ch/site/assets/uploads/2018/02/ipcc_wg3_ar5_annex-iii.pdf#page=7

[386] U.S. National Renewable Energy Laboratory 2021 Update: Lifecycle greenhouse gas emissions from electricity generation. September 2021. https://www.nrel.gov/docs/fy21osti/80580.pdf

[387] United Nations Economic Commission for Europe (March 2022). Lifecycle assessments of electricity generation options. https://unece.org/sed/documents/2021/10/reports/life-cycle-assessment-electricity-generation-options

[388] U.S. EIA: Electricity data. https://www.eia.gov/international/data/world/electricity/electricity-generation

[389] Energy Institute. 2023 statistical review of world energy. https://www.energyinst.org/statistical-review

[390] Climate Watch: Historical GHG emissions. https://www.climatewatchdata.org/ghg-emissions?breakBy=regions&end_year=2018&gases=ch4®ions=WORLD%2CWORLD§ors=electricity-heat&start_year=1990

[391] IEA Transport: Sectoral overview (2022). https://www.iea.org/reports/transport

[392] World Resources Institute: Sector by sector. Where do global greenhouse gas emissions come from? https://ourworldindata.org/ghg-emissions-by-sector

[393] International Energy Agency. Transport. Improving the sustainability of passenger and freight transport. https://www.iea.org/topics/transport

[394] U.S. EIA. International Energy Outlook 2016. Transportation sector energy consumption. https://www.eia.gov/outlooks/ieo/pdf/transportation.pdf

[395] The typical term used for this category is light duty vehicles.

[396] Other examples of electric vehicles are hybrid electric vehicles, plug-in hybrids, and fuel cell vehicles.

[397] The environmental performance of electric vehicles depends on the carbon intensity of electricity.

[398] IEA Electric Vehicles (2023). Tracking Report. Energy- electric vehicles avoid oil consumption. https://www.iea.org/reports/electric-vehicles

[399] U.S. EIA (2023). Global Oil markets. https://www.eia.gov/outlooks/steo/report/global_oil.php

[400] The global fleet of electric vehicles decreased oil (liquid fuel) consumption by 0.70 million barrels per day in 2022. The global EV fleet increased the electricity use by 110 TWh. https://www.iea.org/reports/electric-vehicles 1 gallon gasoline releases 8887 grams of greenhouse gases (GHG). https://www.epa.gov/greenvehicles/greenhouse-gas-emissions-typical-passenger-vehicle The average carbon intensity of electrical grid was 436 grams of GHG per kWh in 2022. https://ember-climate.org/insights/research/global-electricity-review-2023/ The above data was used to estimate the net reduction in GHG for the year 2022 by global EVs. The decrease in GHG from a lower liquid fuel use was 95 million tons. The GHG produced from the extra electricity use was 48 million tons. The net reduction of GHG by the global EVs was 47 million tons in the year 2022. The global GHG emissions are around 50 billion tons per year. Thus, the use of the global electric vehicle fleet reduced GHG by less than 0.1% in 2022.

[401] Relative to coal power, natural gas power has 50% lower GHG emissions per kWh of electricity generated. The global coal power plants emit 10.5 billion tons of GHG per year. https://www.iea.org/reports/global-energy-review-CO_2-emissions-in-2021-2. 1% of 10.5 billion GHG tons is 105 million GHG tons. This means that replacing 1% coal power with natural gas power will reduce GHG by 52 million tons per year. From the previous reference, the impact of electric vehicles is a yearly reduction of 47 million tons. What about replacing coal power with nuclear power? Only 0.5% of coal power needs to be replaced by nuclear power for reducing 50 million tons of GHG. This is because nuclear power has 95% lower GHG emissions relative to coal power.

[402] World Economic Forum. Supply Chains. https://www.weforum.org/agenda/2016/04/the-number-of-cars-worldwide-is-set-to-double-by-2040

[403] IEA Report. Global EV outlook 2023. https://www.iea.org/reports/global-ev-outlook-2023/trends-in-electric-light-duty-vehicles# There are 26 million electric cars out of total of 1.2 billion cars.

[404] Climate Watch: Historical GHG emissions. https://www.climatewatchdata.org/ghg-emissions?breakBy=regions&end_year=2018&gases=ch4®ions=WORLD%2CWORLD§ors=electricity-heat&start_year=1990

[405] World Resources Institute: Sector by sector. Where do global greenhouse gas emissions come from? https://ourworldindata.org/ghg-emissions-by-sector

[406] IEA. Global CO_2 emission from Transport by sub-sector: 2000 to 2030. https://www.iea.org/data-and-statistics/charts/global-co2-emissions-from-transport-by-subsector-2000-2030

[407] Our World in Data. Carbon Intensity of electricity. World average is 436 gCO_2/kWh. https://ourworldindata.org/grapher/carbon-intensity-electricity?tab=table

[408] IEA: Comparative life cycle GHG emissions from a midsize BEV and ICE vehicle (Updated October 2022). https://www.iea.org/data-and-statistics/charts/comparative-life-cycle-greenhouse-gas-emissions-of-a-mid-size-bev-and-ice-vehicle Electric cars emit 50% fewer greenhouse gases compared to conventional cars when considering global average carbon intensity.

[409] The 50% reduction is too optimistic based on recent studies. Prior studies used the average carbon intensity of the grid. Of more relevance is the carbon intensity of the marginal electricity being used by electric cars. The marginal electricity is typically provided by coal power or natural gas. So, the carbon intensity of the marginal electricity is much higher. This translates to a much lower benefit from electric cars. https://www.pnas.org/doi/abs/10.1073/pnas.2116632119

[410] The global fleet of conventional cars contributes to 7% of the total greenhouse gases. Replacing all conventional cars with electric cars that do not emit any greenhouse gases will reduce the global greenhouse gases by 7%. Replacing conventional cars with electric cars that emit 50% less greenhouse gases will

reduce greenhouse gases by 3.5% very year. Countries will continue to lower the emissions from electricity. What if we consider the emissions over the life of the vehicle? It will not make a notable difference. This is because the decrease in the carbon intensity of electricity will be gradual. Also, the 50% reduction is too optimistic based on recent studies. Prior studies used the average carbon intensity of the grid. The carbon intensity of the marginal electricity being used by electric cars is more relevant. The marginal electricity is typically provided by coal or natural gas. So, the carbon intensity of the marginal electricity is much higher. This translates to a lower benefit from electric cars.
https://www.pnas.org/doi/abs/10.1073/pnas.2116632119

[411] The travel range of the cars should also be comparable. For a fair comparison, the electric car should at-least have a travel range that is no less than 25% shorter.

[412] Prices must be compared for the same trim for a fair comparison. SE trim data is presented in the main text. Kona Electric (SEL trim) is 55% higher than the Kona conventional (SEL trim). Kona Electric (Limited trim) is 44% higher than the Kona conventional (Limited trim). Prices are from the Hyundai retail website for the year 2023 models. Notably, the price comparison is at a lower driving range for electric cars. Kona electric has a driving range of 258 miles. While Kona has a driving range of 422 miles. https://www.fueleconomy.gov/

[413] Ford Lightning F150 XLT (EV) with a travel range of 230 miles is over 30% more expensive than Ford F150 XLT with a travel range of over 460 miles. Ford Lightning F150 XLT (EV) with an extended travel range of 320 miles is over 60% more expensive. Only the extended travel range vehicle affords a reasonable comparison. Ford has lost over 30,000 $ per electric vehicle with this pricing in recent quarters.
https://www.ford.com/trucks/f150/f150-lightning/
https://www.fueleconomy.gov/
https://www.reuters.com/business/autos-transportation/ford-withdraws-2023-forecast-warns-ev-results-2023-10-26/#:~:text=Ford%20lost%20an%20estimated%20%2436%2C000,EV%20in%20the%20second%20quarter.

[414] In this book I use the term hybrid cars for those that do not need external charging. I exclude plug in hybrids.

[415] A direct comparison between hybrid and electric cars is not available for the same trim. But direct comparison for the same trims is available between hybrid and conventional cars. It is also available for electric and conventional cars. This can be used to compare hybrid and electric cars. Hybrid cars for the same trim are 10 to 15% more expensive than conventional cars. Examples: Hyundai Elantra Hybrid (Limited trim) is 10% more expensive than Hyundai Elantra (Limited trim). Hyundai Sonata Hybrid (SEL trim) is 15% more expensive than Hyundai Sonata (SEL trim). Toyota Camry Hybrid (LE trim) is 8% more expensive than Toyota Camry (LE). Prices are from the retailer websites for the year 2023 car models. Electric cars are about 40 to 55% more expensive than conventional cars. See previous reference.

[416] Hybrid cars do not require charging. They use liquid fuels. Hybrid cars have a better energy efficiency than conventional cars. Plug in hybrids do not belong in this category.

[417] IEA. Electric vehicles. Lifetime cost of ownership tool. https://www.iea.org/data-and-statistics/data-tools/electric-vehicles-total-cost-of-ownership-tool I provide examples of lifetime cost data from the IEA site (compiled on 25th March 2023) for electric cars and petrol cars. India: electric car: $27,784, Petrol car: $25,924. Indonesia: electric: $29,984, Petrol: $23,207. Ukraine: electric car: $40,475 Petrol: $35,238. Diesel cars are also superior.

[418] Argonne National Laboratory (April 2021): Comprehensive total cost of ownership calculation for vehicles with different size classes and power trains. https://publications.anl.gov/anlpubs/2021/05/167399.pdf

[419] MIT Energy Initiative, Insights into Future Mobility (November 2019): http://energy.mit.edu/wp-content/uploads/2019/11/Insights-into-Future-Mobility.pdf

[420] Environmental defense fund (April 2022). Electric vehicle market update. https://blogs.edf.org/climate411/wp-content/blogs.dir/7/files/2022/04/electric_vehicle_market_report_v6_april2022.pdf

[421] A higher retail price also leads to higher financing and insurance costs. Electric vehicles also require more frequent tire changes.

[422] U.S. DOE. Find and compare cars. https://www.fueleconomy.gov/feg/findacar.shtml The driving range can be found for most cars on this website. A typical electric car has a travel range of 250 miles. A typical hybrid or conventional car has a travel range of more than 400 miles.

[423] A higher travel range requires a larger battery. Large batteries are expensive. Consider Model 3 Tesla. The long-range model is $7000 more expensive than the base model. The long-range model has a travel range of 333 miles. The base model has a travel range of 272 miles.

[424] National Renewable Energy Laboratory (October 2021) Report. There is no place like home: residential parking, electrical access, and implications for the future of electric vehicle charging infrastructure. https://www.nrel.gov/docs/fy22osti/81065.pdf

[425] The International Council on Clean Transportation White Paper (July 2021): Charging up America: Assessing the growing needs of U.S. infrastructure through 2030. https://theicct.org/publication/charging-up-america-assessing-the-growing-need-for-u-s-charging-infrastructure-through-2030/

[426] U.S. Alternative fuels data center. Alternative fueling station locator https://afdc.energy.gov/stations/#/analyze

[427] The International Council on Clean Transportation White Paper (July 2021): Charging up America: Assessing the growing needs of U.S. infrastructure through 2030. https://theicct.org/publication/charging-up-america-assessing-the-growing-need-for-u-s-charging-infrastructure-through-2030/

[428] U.S. Bureau of Transportation statistics: https://www.bts.gov/content/number-us-aircraft-vehicles-vessels-and-other-conveyances

[429] U.S. Alternative fuels data center. Developing infrastructure to charge plugin electric vehicles. https://afdc.energy.gov/fuels/electricity_infrastructure.html#dc

[430] U.S. Alternative fuels data center. Alternative fueling station locator

https://afdc.energy.gov/stations/#/analyze?country=US&fuel=ELEC

[431] ACEA. E-mobility: Only 1 in 9 charging points in EU is fast. https://www.acea.auto/press-release/e-mobility-only-1-in-9-charging-points-in-eu-is-fast/

[432] U.S. Alternative fuels data center. Developing infrastructure to charge plugin electric vehicles. https://afdc.energy.gov/fuels/electricity_infrastructure.html#dc

[433] Diffusion of Innovations. Rogers Everett. ISBN: 978-0-7432-5823-4 I include the innovators (3.5%) and early adopters (13.5%) in the category of Early Adopters.

[434] For example, logistics become a problem beyond a certain production level. https://corporatefinanceinstitute.com/resources/economics/economies-of-scale/

[435] China has about 100 battery gigafactories. Tesla alone has five battery gigafactories. https://www.fdiintelligence.com/content/feature/how-china-is-charging-ahead-in-the-ev-race-80771 https://www.capitalone.com/cars/learn/finding-the-right-car/what-you-need-to-know-about-teslas-gigafactories/1810

[436] U.S. Alternative fuels data center. Alternative fueling station locator https://afdc.energy.gov/stations/#/analyze?country=US&fuel=ELEC

[437] ACEA. E-mobility: Only 1 in 9 charging points in EU is fast. https://www.acea.auto/press-release/e-mobility-only-1-in-9-charging-points-in-eu-is-fast/

[438] Mach1 Road assistance services. Average costs of using car charging stations. https://www.mach1services.com/costs-of-using-car-charging-stations/

[439] EVgo fast charging. EVgo fast charging pricing. https://www.evgo.com/pricing/

[440] Shell: How much does Shell recharge cost? https://support.shell.com/hc/en-gb/articles/115002988472-How-much-does-Shell-Recharge-cost-

[441] EnelX: The ultimate guide to electric vehicle public charging. https://evcharging.enelx.com/resources/blog/579-the-ultimate-guide-to-electric-vehicle-public-charging-pricing

[442] Regular travel across states or countries.

[443] U.S. Energy Information Administration. Few transportation fuels surpass the energy densities of gasoline and diesel. https://www.eia.gov/todayinenergy/detail.php?id=14451

[444] Tesla trucks have a 300-mile and 500-mile travel range. Tesla Website. Semi. https://www.tesla.com/semi Conventional vehicles have a travel range between 1000 and 2000 miles. A semi has fuel tank(s) capacity of 200 to 300 gallons of diesel fuel and a fuel economy of about 5 to 7 miles/gallon.

[445] Electric vehicles have high energy efficiencies. This offsets some of the impact from the poor energy density of the batteries. But only to a small extent.

[446] Payload can be goods or passengers.

[447] U.S. Environmental Protection Agency. Electric vehicle myths. https://www.epa.gov/greenvehicles/electric-vehicle-myths

[448] IEA Analysis. Comparative life cycle GHG emissions from a midsize BEV and ICE vehicle. https://www.iea.org/data-and-statistics/charts/comparative-life-cycle-greenhouse-gas-emissions-of-a-mid-size-bev-and-ice-vehicle

[449] The International council on clean transportation (ICCT). A global comparison of the lifecycle gas emissions of combustion engine and electric passenger cars. https://theicct.org/sites/default/files/publications/Global-LCA-passenger-cars-jul2021_0.pdf

[450] U.S. Department of Energy report. Lifecycle GHG emissions from small sport utility vehicles. https://www.hydrogen.energy.gov/pdfs/21003-life-cycle-ghg-emissions-small-suvs.pdf

[451] IEA Analysis. Comparative life cycle GHG emissions from a midsize BEV and ICE vehicle. https://www.iea.org/data-and-statistics/charts/comparative-life-cycle-greenhouse-gas-emissions-of-a-mid-size-bev-and-ice-vehicle

[452] The IEA analysis provides greenhouse emissions data for electric vehicles for regions with a low, medium, and high carbon intensity.

[453] The effective carbon intensity can be very high in few regions. This occurs where coal power provides the marginal electricity used in the electric cars.

[454] IEA Analysis. Comparative life cycle GHG emissions from a midsize BEV and ICE vehicle. https://www.iea.org/data-and-statistics/charts/comparative-life-cycle-greenhouse-gas-emissions-of-a-mid-size-bev-and-ice-vehicle This analysis does not consider the impact of marginal intensity of electricity. It uses average intensity of the electrical grid. The correct use of marginal intensity will lower the positive impact from electric cars in many regions.

[455] The 50% reduction is too optimistic based on recent studies. Prior studies used the average carbon intensity of the grid. The carbon intensity of the marginal electricity used by electric cars is more relevant. The marginal electricity is typically provided by coal power or natural gas. So, the carbon intensity of the marginal electricity is typically higher. This translates to a lower benefit from electric cars. https://www.pnas.org/doi/abs/10.1073/pnas.2116632119

[456] IScience. Revisiting electric vehicle life cycle greenhouse gas emissions in China. A marginal emission perspective. Volume 26. Page 106565. Year 2023. https://www.sciencedirect.com/science/article/pii/S2589004223006429

[457] Sustainability. Cradle to grave life cycle analysis of ghg emissions of light duty passenger vehicles in China. Volume 15. Page 2627. Year 2023. https://www.mdpi.com/2071-1050/15/3/2627

[458] U.S. Department of Energy: Alternative Fuels data center. Electric vehicles benefits and considerations. https://afdc.energy.gov/fuels/electricity_benefits.html

[459] Recurrent: How long do electric car batteries last? https://www.recurrentauto.com/research/how-long-do-ev-batteries-last

[460] Volkeswagenag.com: Interview with Frank Blome. https://www.volkswagenag.com/en/news/stories/2019/04/our-batteries-last-the-life-of-a-car.html

[461] The hope is that consumers will be alright with a travel range that degrades with time until it reaches 70%.

[462] U.S. Department of Energy: Alternative Fuels data center. Maintenance and safety of electric vehicles. https://afdc.energy.gov/vehicles/electric_maintenance.html

[463] Car and Driver: Electric car battery life. https://www.caranddriver.com/research/a31875141/electric-car-battery-life/

[464] JD Power: How long do electric batteries last? https://www.jdpower.com/cars/shopping-guides/how-long-do-electric-car-batteries-last

[465] Climate watch: Historical GHG emissions. https://www.climatewatchdata.org/ghg-emissions?breakBy=sector&chartType=percentage&end_year=2018&start_year=1990

[466] It generates over 37 billion tons of GHGs each year. The GHG emissions are expressed in terms of CO_2 equivalent. They include the emissions of other GHGs such as methane and nitrous oxide.

[467] United Nations: Climate action. https://www.un.org/en/climatechange/net-zero-coalition

[468] IEA Updated 2023 report: Net zero by 2050. https://www.iea.org/reports/net-zero-roadmap-a-global-pathway-to-keep-the-15-0c-goal-in-reach IEA commentary. https://www.iea.org/commentaries/reaching-net-zero-emissions-demands-faster-innovation-but-weve-already-come-a-long-way

[469] IEA: United Net zero by 2050. https://www.iea.org/reports/net-zero-by-2050

[470] IRENA: World transitions outlook: 1.5°C pathway. https://irena.org/publications/2021/Jun/World-Energy-Transitions-Outlook

[471] Energies. A brief climatology of dunkelflaute events over and surrounding the North and Baltic Sea areas. Vol. 14, Page 6508, Year 2021. https://doi.org/10.3390/en14206508

[472] U.S. EIA: Electricity. Hourly electricity grid monitor. https://www.eia.gov/electricity/gridmonitor/dashboard/electric_overview/US48/US48.

[473] Consider the performance of solar and wind power over the past year in the United States as an example. There were consecutive days (two or more) where the combined output from solar and wind was drastically reduced in all its grids. This was also true for the whole of the United States. Even if all the grids in the U.S. were connected, there would still be two or more consecutive days that have a very low combined output from solar and wind.

[474] IEA: United Net zero by 2050. https://www.iea.org/reports/net-zero-by-2050

[475] Currently, the project cost of solar and wind power plant with just 12 hours of battery storage is five times more than the cost of a natural gas power plant to provide the same amount of annual electricity. The practical energy storage needs for most regions will be 24+ hours to many days. Thus, the project costs with current technology will be extremely high. Data for costs is from the following references. https://www.eia.gov/outlooks/aeo/assumptions/pdf/elec_cost_perf.pdf https://www.nrel.gov/docs/fy23osti/85332.pdf Green hydrogen is even more costly.

[476] U.S. Department of Energy. The pathway to long duration energy storage commercial lift off. https://liftoff.energy.gov/long-duration-energy-storage/

[477] International Energy Agency: Improving the sustainability of passenger and freight transport. https://www.iea.org/topics/transport

[478] U.S. EIA. Few transportation fuels surpass the energy densities of gasoline and diesel. https://www.eia.gov/todayinenergy/detail.php?id=9991

[479] U.S. DOE Report. Quadrennial technology review (2015). Innovative clean technologies in advanced manufacturing. https://www.energy.gov/sites/prod/files/2016/06/f32/QTR2015-6I-Process-Heating.pdf

[480] McKinsey & Company Report. Plugging in. What electrification can do for industry? https://www.mckinsey.com/~/media/McKinsey/Industries/Electric%20Power%20and%20Natural%20Gas/Our%20Insights/Plugging%20in%20What%20electrification%20can%20do%20for%20i

ndustry/Plugging-in-What-electrification-can-do-for-industry-vF.pdf

[481] IEA Report. Net zero Roadmap 2023 update (Figure 1.23). https://www.iea.org/reports/net-zero-roadmap-a-global-pathway-to-keep-the-15-0c-goal-in-reach

[482] About 35 to 40% are still in the demonstration phase. Historically new energy technologies have required several decades to move from prototype/demonstration phase to a wide uptake. What about the ones already commercially available? Even the popular technologies still require large subsidies for consumer uptake. They are not yet competitive in many regions. These technologies suffer from a high upfront cost or convenience issues. Examples are electric cars and heat pumps. The market uptake for these technologies has been mostly driven by large subsidies. The fact that these technologies require large subsidies tells us that they are not yet competitive.

[483] U.S. EIA: Electricity data. https://www.eia.gov/international/data/world/electricity/electricity-generation

[484] International Energy Agency. Renewable electricity. https://www.iea.org/reports/renewable-electricity

[485] IEA: United Net zero by 2050. https://www.iea.org/reports/net-zero-by-2050

[486] IEA Report. Net zero Roadmap 2023 update. https://www.iea.org/reports/net-zero-roadmap-a-global-pathway-to-keep-the-15-0c-goal-in-reach

[487] Many of the commercially available low carbon technologies are not yet competitive. The rapid market uptake of these technologies has mostly been driven by massive subsidies.

[488] AZQuotes.com. Quote has been attributed to Albert Einstein. It is difficult to ascertain if this quote was indeed from Einstein. In any case, it is a wise quote. https://www.azquotes.com/author/4399-Albert_Einstein/tag/assumption

[489] The lifetime cost includes upfront cost and other costs incurred during the life of the technology. Examples include operations & maintenance, and finance costs.

[490] The lifetime cost of technologies that use electricity is comparable or lower only in some countries. A substantial rise in electricity cost will make them higher cost options in all countries. Only a very few applications might have a lower lifetime cost in a few regions.

[491] I will discuss the examples of electric cars. Electric cars are about 40 to 50% more expensive than conventional cars. They have a higher lifetime cost in many countries even at current electricity prices. They are close to break-even in the other countries. A substantial rise in electricity cost will make them a higher cost options in all countries.

[492] National Science Foundation: Science & Engineering Indicators. https://ncses.nsf.gov/pubs/nsb20214/publication-output-by-country-region-or-economy-and-scientific-field

[493] Google Scholar: Use the search term "scientific breakthroughs". https://scholar.google.com

[494] Only a very tiny fraction of breakthroughs will qualify as high-impact breakthroughs. Start-ups also often tout breakthroughs.

[495] Long term often means 10 to 30 years in the energy industry.

[496] Nobody wants an inferior energy experience. Only government policies can drive a wide acceptance of an inferior product. Such policies are likely to be very unpopular.

[497] Photovoltaics report. Fraunhofer Institute for solar energy systems (2020). https://www.ise.fraunhofer.de/content/dam/ise/de/documents/publications/studies/Photovoltaics-Report.pdf

[498] Solar power cost has declined rapidly over the last decade. A high-impact breakthrough–such as the discovery of a superior new material–has not caused the decline. The cost decline has mainly been the result of a rapid increase in the production of silicon-based technology. An increase in production levels is associated with an increase in efficiency because of the benefits from economy-of-scale. Over time, there have been many incremental technical improvements in the various steps of the process. This is another major reason for the cost declines.

[499] U.S. Department of Energy: Perovskite solar cells. https://www.energy.gov/eere/solar/perovskite-solar-cells

[500] Greentech media. Rest in peace. List of deceased solar companies https://www.greentechmedia.com/articles/read/rest-in-peace-the-list-of-deceased-solar-companies

[501] Greentech media. Hard lessons from the great algae biofuels bubble. https://www.greentechmedia.com/articles/read/lessons-from-the-great-algae-biofuel-bubble

[502] Biofuels watch. Dead end road. The false promises of cellulosic biofuels. https://www.biofuelwatch.org.uk/wp-content/uploads/Cellulosic-biofuels-report-low-resolution.pdf

[503] Recall, a high-impact breakthrough is crucial for wide acceptance of a new technology.

[504] U.S. EIA: Primary energy. https://www.eia.gov/international/data/world/total-energy/total-energy-consumption?

[505] International Energy Agency Report, 2021. Net zero by 2050: A roadmap for the energy sector. https://www.iea.org/reports/net-zero-by-2050

[506] International Renewable Energy Agency. World Energy transitions outlook: 1.5°C pathway. https://www.irena.org/publications/2021/Jun/World-Energy-Transitions-Outlook

[507] U.S. EIA: Primary energy. https://www.eia.gov/international/data/world/total-energy/total-energy-consumption?

[508] Energy Institute. 2023 Statistical review of world energy. https://www.energyinst.org/statistical-review

[509] Our World in Data. Global direct primary consumption. https://ourworldindata.org/grapher/global-primary-energy?country=~OWID_WRL

[510] The mid-point of the last transition was around the year 1900.

[511] Our World in Data. Global direct primary consumption. https://ourworldindata.org/grapher/global-primary-energy?country=~OWID_WRL

[512] Eurostat: Statistics Explained. https://ec.europa.eu/eurostat/statistics-explained/index.php?title=Glossary:Tonnes_of_oil_equivalent_(toe)

[513] IEA: Key World Energy Statistics 2020. https://www.iea.org/reports/key-world-energy-statistics-2020

[514] U.S. EIA: Primary energy. https://www.eia.gov/international/data/world/total-energy/total-energy-consumption?

[515] OECD iLibrary: Global plastics outlook. https://www.oecd-ilibrary.org/sites/de747aef-en/index.html?itemId=/content/publication/de747aef-en The current production of plastics is 500 million tons per year.

[516] World Steel Association: world steel in figures 2022. https://worldsteel.org/steel-topics/statistics/world-steel-in-figures-2022/ The current production of steel is 1900 million tons per year.

[517] IEA: Cement (Activity section) https://www.iea.org/reports/cement The current production of cement is 4300 million tons per year.

[518] U.S. EIA: Primary energy. https://www.eia.gov/international/data/world/total-energy/total-energy-consumption?

[519] United nations. Food and Agriculture organization. Land use in agriculture by numbers. https://www.fao.org/sustainability/news/detail/en/c/1274219/

[520] The following data were used for the estimate. According to the United Nations, the world consumes on an average 2960 kcal per person per day. https://news.un.org/en/story/2022/12/1131637 The global population is 8.1 billion. According to IEA world final energy consumption is over 400 EJ/year. https://www.iea.org/reports/key-world-energy-statistics-2021/final-consumption

[521] IEA Updated 2023 report: Net zero by 2050. https://www.iea.org/reports/net-zero-roadmap-a-global-pathway-to-keep-the-15-0c-goal-in-reach

[522] International Energy Agency Report, 2021. Net zero by 2050: A roadmap for the energy sector. https://www.iea.org/reports/net-zero-by-2050

[523] U.S. EIA: Electricity data. https://www.eia.gov/international/data/world/electricity/electricity-generation

[524] Our World in Data. Electricity generation. https://ourworldindata.org/grapher/electricity-generation?tab=chart&country=~OWID_WRL

[525] Ember climate. Global electricity review 2023. https://ember-climate.org/insights/research/global-electricity-review-2023/#supporting-material

[526] Our World in Data. Electricity generation. https://ourworldindata.org/grapher/electricity-generation?tab=chart&country=~OWID_WRL

[527] IEA data. Global hydrogen review 2022. https://www.iea.org/reports/global-hydrogen-review-2022/executive-summary

[528] Clean vehicles are those with zero tailpipe emissions. Examples are battery electric vehicles and hydrogen fuel cell electric vehicles. Battery electric vehicles dominate this category.

[529] IEA data. Trends in charging infrastructure. https://www.iea.org/reports/global-ev-outlook-2023/trends-in-charging-infrastructure

[530] The IEA report distinguished new building from existing buildings. I have included both the categories.

[531] IEA Report, revised March 2022. The role of critical minerals on clean energy transitions. https://www.iea.org/reports/the-role-of-critical-minerals-in-clean-energy-transitions

[532] World Nuclear Association. Minerals requirement for electricity generation. https://www.world-nuclear.org/information-library/energy-and-the-environment/mineral-requirements-for-electricity-generation.aspx

[533] IEA Report, 2021. The role of critical minerals on clean energy transitions. See next reference for details. https://www.iea.org/reports/the-role-of-critical-minerals-in-clean-energy-transitions

[534] IEA provides data based on capacity (per MW). But the capacity factor or availability of wind and solar power to generate electricity is lower than natural gas and coal power. Also, the life of solar and wind power is lower than natural gas and coal power. So, a solar or wind power plant with the same capacity as a fossil fuel power plant produces much lower electricity over its lifetime. A more relevant approach is to compare the materials

required for the electricity produced by the power plants over the lifetime. I have converted the IEA data from materials required per capacity to materials required per unit of electricity generated over the lifetime. I have used average data from IEA reports for the capacity factors and lifetimes. I have used the average of natural gas and coal power to represent fossil fuel power. https://www.iea.org/data-and-statistics/charts/average-annual-capacity-factors-by-technology-2018 https://www.iea.org/reports/energy-technology-perspectives-2023

[535] U.S. Geological Survey. Critical Minerals. https://www.usgs.gov/science/critical-minerals

[536] IEA Report, 2021. The role of critical minerals on clean energy transitions. https://www.iea.org/reports/the-role-of-critical-minerals-in-clean-energy-transitions

[537] A Report by the White House. Building resilient supply chains, revitalizing American manufacturing, and fostering broad based growth. https://www.whitehouse.gov/wp-content/uploads/2021/06/100-day-supply-chain-review-report.pdf

[538] The World Bank Group (2020 report). Minerals for climate action: the minerals intensity of clean energy transition. https://pubdocs.worldbank.org/en/961711588875536384/Minerals-for-Climate-Action-The-Mineral-Intensity-of-the-Clean-Energy-Transition.pdf

[539] IEA Energy Technology Perspectives 2023. https://www.iea.org/reports/energy-technology-perspectives-2023#

[540] IEA Report. Critical minerals market review 2023. https://www.iea.org/reports/critical-minerals-market-review-2023/implications

[541] USGS Copper data: https://pubs.usgs.gov/periodicals/mcs2023/mcs2023-copper.pdf The total global copper production was 21,200 kt in 2021.

[542] USGS Lithium data: https://pubs.usgs.gov/periodicals/mcs2023/mcs2023-lithium.pdf

[543] USGS Nickel data: https://pubs.usgs.gov/periodicals/mcs2023/mcs2023-nickel.pdf

[544] USGS Cobalt data: https://pubs.usgs.gov/periodicals/mcs2023/mcs2023-cobalt.pdf

[545] A workforce with specific skills is required for extracting and processing minerals, fabricating components, building electricity networks, and installing equipment.

[546] IEA Updated 2023 report: Net zero by 2050. https://www.iea.org/reports/net-zero-roadmap-a-global-pathway-to-keep-the-15-0c-goal-in-reach

[547] Many of the commercially available low carbon technologies are not yet competitive. The market uptake of these technologies has mostly been driven by massive subsidies.

[548] The environmental issues include health and safety.

[549] American Chemical Society, Chlorofluorocarbons, and ozone depletion. https://www.acs.org/content/acs/en/education/whatischemistry/landmarks/cfcs-ozone.html

[550] The Washington Post. https://www.washingtonpost.com/archive/politics/1988/04/10/cfcs-rise-and-fall-of-chemical-miracle/9dc7f67b-8ba9-4e11-b247-a36337d5a87b/

[551] US Department of State. https://www.state.gov/key-topics-office-of-environmental-quality-and-transboundary-issues/the-montreal-protocol-on-substances-that-deplete-the-ozone-layer/

[552] UN Environment program. https://ozone.unep.org/ozone-timeline

[553] U.S. Environmental Protection Energy, https://www.epa.gov/ozone-layer-protection/international-actions-montreal-protocol-substances-deplete-ozone-layer

[554] Our World in Data. Plastic Pollution. https://ourworldindata.org/plastic-pollution

[555] UN Environment Program. Global Chemicals Outlook II, From Legacies to Innovative solutions (2019). https://wedocs.unep.org/bitstream/handle/20.500.11822/27651/GCOII_synth.pdf?sequence=1&isAllowed=y

[556] UN Environment program. The state of plastics. World Environment Day outlook (2018). https://wedocs.unep.org/bitstream/handle/20.500.11822/25513/state_plastics_WED.pdf?sequence=1&isAllowed=y

[557] UN Environment. The state of plastics. World Environment Day outlook (2018). https://wedocs.unep.org/bitstream/handle/20.500.11822/25513/state_plastics_WED.pdf?sequence=1&isAllowed=y

[558] Directive 2003/30/EC of the European Parliament on the promotion of biofuels or other renewable fuels for transport. https://eur-lex.europa.eu/legal-content/EN/TXT/?qid=1586459020405&uri=CELEX:32003L0030

[559] Technical report for EU commission. Study on the environmental impact of palm oil consumption. https://op.europa.eu/en/publication-detail/-/publication/89c7f3d8-2bf3-11e8-b5fe-01aa75ed71a1

[560] European Parliament resolution of 4 April 2017 on palm oil and deforestation of rainforests. https://www.europarl.europa.eu/doceo/document/TA-8-2017-0098_EN.html?redirect

[561] J. R. Fleming, Eos, Transactions American Geophysical Union 79, 405 (1998). https://agupubs.onlinelibrary.wiley.com/doi/abs/10.1029/98EO00310

[562] E. Hawkins, P.D. Jones, Quarterly Journal of the Royal Meteorological Society, 139, 1961 (2013). https://www.researchgate.net/publication/251231528_On_increasing_global_temperatures_75_years_after_Callendar https://centaur.reading.ac.uk/32981/1/hawkins_jones_2013.pdf

[563] United Nations: From Stockholm to Kyoto- A brief history of climate change. https://www.un.org/en/chronicle/article/stockholm-kyoto-brief-history-climate-change#

[564] American Institute of Physics. A history of global warming. https://history.aip.org/climate/summary.htm https://history.aip.org/climate/CO2.htm

[565] A significant concern about the climate impacts from fossil fuels only emerged in the 1970s. https://history.aip.org/climate/CO2.htm

[566] Our World in Data. Global direct energy consumption. https://ourworldindata.org/grapher/global-primary-energy?country=~OWID_WRL

[567] The data can be obtained from following two references. "Carbon Dioxide Information Analysis Center". T. Boden, D. Andres, Oakridge National Laboratory. https://cdiac.ess-dive.lbl.gov/ftp/ndp030/global.1751_2014.ems and BP Statistical review of World energy 2021. https://www.bp.com/content/dam/bp/business-sites/en/global/corporate/pdfs/energy-economics/statistical-review/bp-stats-review-2021-full-report.pdf

[568] E. Hawkins, P.D. Jones, Quarterly Journal of the Royal Meteorological Society, 139, 1961 (2013). https://www.researchgate.net/publication/251231528_On_increasing_global_temperatures_75_years_after_Callendar https://centaur.reading.ac.uk/32981/1/hawkins_jones_2013.pdf

[569] International Energy Agency Report, 2021. Net zero by 2050: A roadmap for the energy sector. https://www.iea.org/reports/net-zero-by-2050

[570] International Renewable Energy Agency. World Energy transitions outlook: 1.5°C pathway. https://www.irena.org/publications/2021/Jun/World-Energy-Transitions-Outlook

[571] Help from nature reduces the challenge of the task and thereby lowers the resource intensity. For example, less effort is required to swim against the water current as opposed to with the water current.

[572] OECD Report. Material resources, productivity and the environment. https://www.oecd.org/greengrowth/MATERIAL%20RESOURCES,%20PRODUCTIVITY%20AND%20THE%20ENVIRONMENT_key%20findings.pdf

[573] European Environment Agency. Resource use and materials. https://www.eea.europa.eu/en/topics/in-depth/resource-use-and-materials

[574] German Environment Agency (Umwelt Bundesamt). Resource use and its consequences.

https://www.umweltbundesamt.de/en/topics/waste-resources/resource-use-its-consequences#

[575] There are different types of resources. Solar and wind power require more critical minerals and land. While fossil fuel power plants require very large amount of fossil fuels. The disparate resources cannot be mathematically added. My approach, discussed in the text, allows the comparison of the overall level of resources required.

[576] IEA Energy Technology Perspectives 2023. https://www.iea.org/reports/energy-technology-perspectives-2023#

[577] IEA provides data based on capacity (per MW). But the capacity factor or availability of wind and solar power to generate electricity is lower than natural gas and coal power. Also, the life of solar and wind power is lower than natural gas and coal power. So, a solar or wind power plant with the same capacity as a fossil fuel power plant produces much lower electricity over the life. A more relevant approach is to compare the materials required per unit electricity produced by the power plants over the lifetime. I have converted the IEA data from materials required per capacity to materials required per unit electricity generated over the lifetime. I have used average data from IEA reports for the capacity factors and lifetimes. I have used the average of natural gas and coal power to represent fossil fuel power. https://www.iea.org/data-and-statistics/charts/average-annual-capacity-factors-by-technology-2018 https://www.iea.org/reports/energy-technology-perspectives-2023

[578] IEA Report, revised March 2022. The role of critical minerals on clean energy transitions. https://www.iea.org/reports/the-role-of-critical-minerals-in-clean-energy-transitions

[579] World Nuclear Association. Mineral requirement for electricity generation. https://www.world-nuclear.org/information-library/energy-and-the-environment/mineral-requirements-for-electricity-generation.aspx

[580] Oakridge National Laboratory Report. Environmental quality and U.S. power sector. Air quality, water quality, land use and environmental justice.

https://www.energy.gov/sites/prod/files/2017/01/f34/Environment%20Baseline%20Vol.%202--Environmental%20Quality%20and%20the%20U.S.%20Power%20Sector--Air%20Quality%2C%20Water%20Quality%2C%20Land%20Use%2C%20and%20Environmental%20Justice.pdf

[581] IEA Energy Technology Perspectives 2023. https://www.iea.org/reports/energy-technology-perspectives-2023#

[582] IEA provides data based on capacity (per MW). But the capacity factor or availability of wind and solar power to generate electricity is lower than natural gas and coal power. Also, the life of solar and wind power is lower than natural gas and coal power. So, a solar or wind power plant with the same capacity as a fossil fuel power plant produces much lower electricity over the life. A more relevant approach is to compare the materials required per unit electricity produced by the power plants over the lifetime. I have converted the IEA data from materials required per capacity to materials required per unit electricity generated over the lifetime. I have used average data from IEA reports for the capacity factors and lifetimes. I have used the average of natural gas and coal power to represent fossil fuel power. https://www.iea.org/data-and-statistics/charts/average-annual-capacity-factors-by-technology-2018 https://www.iea.org/reports/energy-technology-perspectives-2023

[583] IEA Energy Technology Perspectives 2023. https://www.iea.org/reports/energy-technology-perspectives-2023#

[584] Environmental Science and technology. Rock to metal ratio. A foundational metric for understanding my wastes. Volume 56. Page 6710. Year 2022. https://pubs.acs.org/doi/10.1021/acs.est.1c07875

[585] The rock to metal ratio is 9 and 7 (tons/ton) for iron and aluminum, respectively. I have used their average (8 tons/ton) for this estimate. The data is from IEA technology perspectives.

[586] The water is 0.6 and 0.4 (m^3/ton) for iron and aluminum, respectively. I have used their average (0.5 m^3/ton) for this estimate. The data is from IEA technology perspectives.

[587] IEA Report, revised March 2022. The role of critical minerals on clean energy transitions. https://www.iea.org/reports/the-role-of-critical-minerals-in-clean-energy-transitions

[588] MIT Climate portal. Will mining the resources needed for clean energy cause problems for the environment? https://climate.mit.edu/ask-mit/will-mining-resources-needed-clean-energy-cause-problems-environment

[589] Materialstoday Proceedings. Life cycle assessment of electric vehicles in comparison to combustion engine vehicles. A review. Volume 49. Page 217. Year 2022. https://www.sciencedirect.com/science/article/abs/pii/S221478532100763X

[590] Science of the total environment. A review of the life cycle assessment of electric vehicles. Volume 814. Page 152870. Year 2022. https://www.sciencedirect.com/science/article/abs/pii/S0048969721079493

[591] EIA International data. Petroleum and Other liquids. https://www.eia.gov/international/data/world/petroleum-and-other-liquids/annual-petroleum-and-other-liquids-production?

[592] EIA International data. Natural gas. https://www.eia.gov/international/data/world/natural-gas/dry-natural-gas-production?

[593] EIA International data. Coal and Coke. https://www.eia.gov/international/data/world/coal-and-coke/coal-and-coke-production?

[594] IEA Coal Consumption. https://www.iea.org/reports/coal-information-overview/consumption

[595] EIA International data. Petroleum and Other liquids. https://www.eia.gov/international/data/world/petroleum-and-other-liquids/annual-petroleum-and-other-liquids-production?

[596] EIA International data. Natural gas. https://www.eia.gov/international/data/world/natural-gas/dry-natural-gas-production?

[597] EIA International data. Coal and Coke. https://www.eia.gov/international/data/world/coal-and-coke/coal-and-coke-production?

[598] IEA Report. Critical minerals market review 2023. https://www.iea.org/reports/critical-minerals-market-review-2023/implications

[599] IEA Energy Technology Perspectives 2023. https://www.iea.org/reports/energy-technology-perspectives-2023#

[600] International Energy Agency Report, 2021. Net zero by 2050: A roadmap for the energy sector. https://www.iea.org/reports/net-zero-by-2050

[601] Bureau of International Recycling. World recycling in figures 2017-2021. https://www.bir.org/images/BIR-pdf/Ferrous_report_2017-2021_lr.pdf

[602] International Aluminum. Recycling Factsheet. https://international-aluminium.org/resource/aluminium-recycling-fact-sheet/#:~:text=According%20to%20the%20data%2C%20the,recycling%20efficiency%20rate%20is%2076%25.

[603] United National Environment Programme. International resource panel. Recycling rates of metals. https://www.resourcepanel.org/reports/recycling-rates-metals

[604] United National Environment Programme. Time to seize opportunity, tackle challenge of e-waste. https://www.unep.org/news-and-stories/press-release/un-report-time-seize-opportunity-tackle-challenge-e-waste

[605] OECD.org Plastic pollution is growing relentlessly as waste management and recycling falls short, says OECD. https://www.oecd.org/environment/plastic-pollution-is-growing-relentlessly-as-waste-management-and-recycling-fall-short.htm

[606] U.S. EPA. Facts and figures about materials, waste, and recycling. https://www.epa.gov/facts-and-figures-about-materials-waste-and-recycling/plastics-material-specific-data

[607] Greenpeace, 2022. Plastic recycling is a dead-end street. https://www.greenpeace.org/usa/news/new-greenpeace-report-plastic-recycling-is-a-dead-end-street-year-after-year-plastic-recycling-declines-even-as-plastic-waste-increases/

[608] MIT Technology Review, Aug. 2021. Solar panels are a pain to recycle. These companies are trying to fix that. https://www.technologyreview.com/2021/08/19/1032215/solar-panels-recycling/

[609] U.S. EIA, Wind Energy and the Environment. https://www.eia.gov/energyexplained/wind/wind-energy-and-the-environment.php

[610] IEA Wind Task 45. Enabling the recycling of wind turbine blades. https://iea-wind.org/task45/

[611] The World Bank Group (2020 report). Minerals for climate action: the minerals intensity of clean energy transition. https://pubdocs.worldbank.org/en/961711588875536384/Minerals-for-Climate-Action-The-Mineral-Intensity-of-the-Clean-Energy-Transition.pdf

[612] IEA. https://www.iea.org/commentaries/reaching-net-zero-emissions-demands-faster-innovation-but-weve-already-come-a-long-way

[613] Even a very popular technology such as electric vehicles has high upfront costs and convenience issues. The problem can be best seen from the fact that a major fraction of the population is still hesitant to use electric cars. https://epic.uchicago.edu/news/4-in-10-say-next-vehicle-will-be-electric-ap-norc-epic-poll/ This is despite the large national and state subsidies.

[614] Such knowledge applies to all energy projects. This is because the practical issues are common to all the large projects in the energy industry.

[615] Major projects have a budget of more than hundred million dollars. They require very large resources over a short timeframe. This is why major projects are very complex.

[616] Developing energy expertise is a slow process. It takes a lot of time to gain access to the required experience. It takes more than ten years for an energy technology to move from the idea phase to commercial application. Developing expertise in technology evaluation requires at-least five years. Major projects take more than five years from initiation to completion.

[617] AZQuote. Albert Einstein Quotes. https://www.azquotes.com/quote/1358685

[618] See the list of authors of the IPCC reports and the authors of the publications cited in the reports. https://apps.ipcc.ch/report/authors/ A large majority only have an academic knowledge about energy. They do not have the relevant experience to qualify as an energy expert. Specifically, they do not have experience with evaluating, developing, and deploying technologies and major projects. Such experience can only be gained via a career in the energy industry.

[619] The authors are capable people. But they are not energy experts because they do not have the required skills, and experience. See the previous reference for details.

[620] Most IEA personnel do not have adequate work experience in the energy industry.
They have limited energy expertise. They have the expertise to obtain quality historical data and discuss future requirements. But they have limitations because of their lack of relevant experience. For example, they do not have the expertise to predict the time required to overcome the technical hurdles associated with the energy transition.

[621] Representation from the energy industry is grossly inadequate. Especially, from the fossil fuel industry. Fossil fuels have contributed to about 80% of the global energy for many decades. Also, the experience of this industry is not limited to fossil fuels. Oil companies have spent billions of dollars on low carbon energy R&D over the decades. Many from the fossil fuel industry have experience with developing low carbon technologies. They have relevant knowledge and experience. They should have a strong representation in the path forward discussions. Yet they have a very poor representation.

[622] European Commission Report. Study on energy subsidies and other government interventions in the European Union. 2022 Edition. https://op.europa.eu/en/publication-detail/-/publication/34a55767-55a1-11ed-92ed-01aa75ed71a1

[623] Energy Institute. 2023 Statistical review of world energy. https://www.energyinst.org/statistical-review

[624] U.S. EIA Report (2023). Federal financial interventions and subsidies in energy in fiscal years 2016-2022. https://www.eia.gov/analysis/requests/subsidy/

[625] Energy Institute. 2023 Statistical review of world energy. https://www.energyinst.org/statistical-review

[626] Details are provided in an earlier section.

[627] How the world really works (2022). Book by Vaclav Smil.

[628] U.S. Department of energy. Products made from oil and natural gas. https://www.energy.gov/sites/prod/files/2019/11/f68/Products%20Made%20From%20Oil%20and%20Natural%20Gas%20Infographic.pdf

[629] High income countries have a higher energy use per capita compared to the global average. Low-income countries have a lower energy use per person. Fossil fuels have been the major energy source globally. See figure in the next chapter.

[630] Energy Institute. 2023 statistical review of world energy. https://www.energyinst.org/statistical-review

[631] IEA Updated 2023 report: Net zero by 2050. https://www.iea.org/reports/net-zero-roadmap-a-global-pathway-to-keep-the-15-0c-goal-in-reach

[632] Many of the commercially available low carbon technologies are not yet competitive. The rapid market uptake of these technologies has mostly been driven by massive subsidies.

[633] A large penetration of solar and wind power requires long duration energy storage. Electric vehicles have a much higher upfront price and require longer wait times to recharge.

[634] An alternative is competitive if it has a performance and a cost that is not substantially inferior to the incumbent.

[635] But a pure free-market driven system does not exist. Governments have introduced regulations to ensure that the system does not go off the rails.

[636] UN Environment Programme. The United States clean air act turns 50. https://www.unep.org/news-and-stories/story/united-states-clean-air-act-turns-50-air-any-better-half-century-later

[637] US EPA. Clean air act requirements and history. https://www.epa.gov/clean-air-act-overview/clean-air-act-requirements-and-history

[638] US EPA. Progress cleaning the air and improving people's health. https://www.epa.gov/clean-air-act-overview/clean-air-act-requirements-and-history

[639] The involvement from governments should be minimized. One approach is to replace subsidies by a smartly designed carbon tax. This should drive the transition while allowing the market to pick the winners. A smartly designed carbon tax would ensure robust prioritization which I discuss in a later chapter. Subsidies should be provided only to support early adoption of technologies. Subsidies should not be provided to technologies that have achieved more than 1% of the market share.

[640] IPCC Special Report: Global warming of 1.5 °C. Chapter 2. Figure 2.3 https://www.ipcc.ch/sr15/chapter/chapter-2/

[641] The White House. Inflation Reduction Act guidebook. https://www.whitehouse.gov/cleanenergy/inflation-reduction-act-guidebook/

[642] Rhodium Group Report. A turning point for U.S. Climate progress. https://rhg.com/research/climate-clean-energy-inflation-reduction-act/

[643] U.S. EIA: Primary energy. https://www.eia.gov/international/data/world/total-energy/total-energy-consumption?

[644] McKinsey & Company Special Report, January 2022. The net zero transition. What it would cost? What it could bring? https://www.mckinsey.com/business-functions/sustainability/our-insights/the-economic-transformation-what-would-change-in-the-net-zero-transition

[645] International Renewable Energy Agency. World Energy transitions outlook: 1.5°C pathway. https://www.irena.org/publications/2021/Jun/World-Energy-Transitions-Outlook

[646] The scope of the policies will depend on the issues faced by that specific region. Examples include the economic state, politics, energy use per person, energy mix, and population of the region.

[647] PBL Netherlands Environmental Assessment Agency 2021 Report. Trends in global CO_2 and GHG emissions. https://www.pbl.nl/en/trends-in-global-CO2-emissions

[648] United Kingdom is considered as a part of EU countries for this discussion.

[649] PBL Netherlands Environmental Assessment Agency 2021 Report. Trends in global CO_2 and GHG emissions. https://www.pbl.nl/en/trends-in-global-co2-emissions

[650] It includes data from the earliest year available. CO_2 is used as a proxy for the greenhouse gases because it is the major component and historical CO_2 data is available.

[651] Global carbon budget, 2022. https://www.icos-cp.eu/science-and-impact/global-carbon-budget/2022

[652] Global carbon budget, 2022. https://www.icos-cp.eu/science-and-impact/global-carbon-budget/2022

[653] Google scholar. Search term "Climate Science". https://scholar.google.com/scholar?hl=en&as_sdt=0%2C44&q=climate+science&btnG=

[654] World Meteorological Organization. https://public.wmo.int/en/our-mandate/what-we-do/observations

[655] National oceanic and atmospheric administration. How do scientists study ancient climates? https://www.ncei.noaa.gov/news/how-do-scientists-study-ancient-climates#:~:text=To%20extend%20those%20records%2C%20paleoclimatologists,in%20the%20rings%20of%20trees.

[656] NASA Global climate change. Global Temperature. https://climate.nasa.gov/vital-signs/global-temperature/

[657] European Environment Agency. Trends in atmospheric concentrations of CO_2, CH_4 and N_2O. https://www.eea.europa.eu/data-and-maps/daviz/atmospheric-concentration-of-carbon-dioxide/#tab-chart_1

[658] NOAA Global Monitoring Laboratory. The Basics. Isotopic Fingerprints. https://gml.noaa.gov/outreach/isotopes/

[659] IPCC Reports. https://www.ipcc.ch/reports/

[660] IPCC AR6. Climate Change 2021. The Physical Science Basis. https://www.ipcc.ch/report/ar6/wg1/downloads/report/IPCC_AR6_WGI_SPM_final.pdf

[661] Climate change is likely causing several other impacts. But scientists at the present time have a low to medium confidence about these impacts because of limited data.

[662] IPCC Reports. https://www.ipcc.ch/reports/

[663] UN Sustainable Development Goals. Goal 6: Clean water and sanitation. https://www.un.org/sustainabledevelopment/water-and-sanitation/

[664] UN Sustainable Development Goals. Goal 2: Zero Hunger. https://www.un.org/sustainabledevelopment/hunger/

[665] UN Sustainable Development Goals. Goal 7: Affordable and clean energy. https://www.un.org/sustainabledevelopment/energy/

[666] UN Sustainable Development Goals. Goal 3: Good health and well-being. https://www.un.org/sustainabledevelopment/health/

[667] UN IGME Report (2022). Levels and trends in child mortality. https://data.unicef.org/resources/levels-and-trends-in-child-mortality/

[668] The entire impact from such events cannot be attributed to climate change. So, the impact is lower than the data suggests.

[669] EM-DAT, CRED/UCLouvain, Brussels, Belgium. CRED Crunch Issues 58,66 & 70 include data from 2019, 2021 and 2022. The report "Human cost of Disasters" includes data from 2000-2019. https://www.emdat.be/publications?field_publication_type_tid=66

[670] Affected people include the injured, homeless, and otherwise affected.

[671] I provide numbers from all climate related disasters. Climate change only contributes to a portion of the impact. Some of the impact would occur even if there was no human contribution. So, the actual impact is smaller than the numbers I discuss here.

[672] International Monetary Fund. General Government Debt. https://www.imf.org/external/datamapper/GG_DEBT_GDP@GDD/

[673] History has shown a very strong correlation between energy access and standard of living. The correlation is very strong until a certain threshold level of energy access is achieved.

[674] Many believe that each year is much worse than the previous in terms of climate impact. And every decade is much worse than the previous decade.

[675] Our World in Data. Natural Disasters. https://ourworldindata.org/natural-disasters The basic data is

from EM-DAT. The data set normalizes for the large increase in population and GDP across the decades.

676 Our World in Data. https://ourworldindata.org/natural-disasters# This site uses the raw data from EM-DAT and normalizes for the change in population. Emergency events database (EM-DAT). The International Disasters Database. https://www.emdat.be UN World population aspects. https://population.un.org/wpp/

677 Our World in Data. https://ourworldindata.org/natural-disasters# This site uses the data from EM-DAT and normalizes for the change in GDP. Emergency events database (EM-DAT). The International Disasters Database. https://www.emdat.be Data from GDP is from World Bank. https://data.worldbank.org

678 IPCC AR6 Report. Climate Change 2021: The Physical Science Basis. https://www.ipcc.ch/report/ar6/wg1/downloads/report/IPCC_AR6_WGI_SPM_final.pdf

679 Our World in Data. Natural Disasters. https://ourworldindata.org/natural-disasters

680 See previous section.

681 UN Environment Programme. Emissions gap report 2023. https://www.unep.org/resources/emissions-gap-report-2023

682 UN Environment Programme. Emissions gap report 2023. Nations must go further than current Paris pledges or face warming of 2.5°C to 2.9°C https://www.unep.org/news-and-stories/press-release/nations-must-go-further-current-paris-pledges-or-face-global-warming#:~:text=Nairobi%2C%2020%20November%202023%20–%20As,pre%2Dindustrial%20levels%20this%20century%2C

683 IEA Report. World energy outlook 2023. https://www.iea.org/reports/world-energy-outlook-2023/executive-summary Stated policies scenario considers existing policies and those that are under development.

684 IEA Report. World energy outlook 2023. https://www.iea.org/reports/world-energy-outlook-2023/executive-summary Announced policies scenario considers both short- and long-term national pledges delivered to meet the climate targets.

685 IPCC Reports. https://www.ipcc.ch/reports/

[686] There is no credible evidence to suggest otherwise. Credible evidence is that based on quality data. Speculative assumptions cannot be used to provide such evidence.

[687] A realistic view is that there is no cliff in the foreseeable future. The impact from climate change will get worse. But humans have endured many challenges. We will manage this challenge as well. We also need a robust reference point. The current impact from climate change is far less than the impact from poverty. The feeling that the impact from climate change is already intolerable is not supported by facts.

[688] GDP per capita of a country is equal to the GDP of the country divided by the total number of people.

[689] Vaclav Smil. Energy and Civilization. A history. MIT Press.

[690] World Bank data. GDP per capita, PPP (current international $). https://data.worldbank.org/indicator/NY.GDP.PCAP.PP.CD

[691] U.S. EIA. International Data. Energy Intensity. https://www.eia.gov/international/data/world/other-statistics/energy-intensity-by-gdp-and-population?

[692] U.S. EIA. International Data. Energy Intensity. https://www.eia.gov/international/data/world/other-statistics/energy-intensity-by-gdp-and-population?

[693] World Bank data. GDP per capita, PPP (current international $). https://data.worldbank.org/indicator/NY.GDP.PCAP.PP.CD Use of GDP data in terms of PPP accounts for the difference in purchasing power of the different countries.

[694] UNDP. Human development reports. https://hdr.undp.org/content/energising-human-development

[695] Our World in data. Human development index vs GDP per capita, 2021. https://ourworldindata.org/grapher/human-development-index-vs-gdp-per-capita

[696] U.S. EIA. International Data. Energy Intensity. https://www.eia.gov/international/data/world/other-statistics/energy-intensity-by-gdp-and-population?

[697] World Bank data. GDP per capita, PPP (current international $). https://data.worldbank.org/indicator/NY.GDP.PCAP.PP.CD

[698] World Bank data. GDP per capita, PPP (current international $). https://data.worldbank.org/indicator/NY.GDP.PCAP.PP.CD

[699] World Bank data. https://pip.worldbank.org/home Data is in terms of 2017 PPP$. It accounts for the difference in the cost of living across countries. I input $10 for the poverty line to get the data.

[700] U.S. EIA report. Cost and performance characteristics of new generation resources in the annual energy outlook 2022. https://www.eia.gov/outlooks/aeo/assumptions/pdf/table_8.2.pdf

[701] U.S. EIA report. Levelized cost of new generation resources in the annual energy outlook 2022. https://www.eia.gov/outlooks/aeo/pdf/electricity_generation.pdf

[702] IEA report. Projected costs of generating electricity. https://iea.blob.core.windows.net/assets/ae17da3d-e8a5-4163-a3ec-2e6fb0b5677d/Projected-Costs-of-Generating-Electricity-2020.pdf

[703] The Energy & Resources Institute. Exploring electricity supply mix scenarios to 2030. https://www.teriin.org/sites/default/files/2019-02/Exploring%20Electricity%20Supply-Mix%20Scenarios%20to%202030.pdf

[704] National Energy Technology Laboratory report. Cost and performance baseline for fossil energy plants. https://www.netl.doe.gov/energy-analysis/details?id=729

[705] National Renewable Energy Laboratory report. U.S. Solar PV and energy storage cost benchmark. https://www.nrel.gov/docs/fy22osti/80694.pdf The cost of solar with just four hours of storage is almost twice higher than solar alone. Long duration storage involves a few to many days of energy storage. The cost of long duration storage will be far higher.

[706] U.S. Energy Information Administration. Cost and performance characteristics of new generation resources in the annual energy outlook 2023. https://www.eia.gov/outlooks/aeo/assumptions/pdf/elec_cost_perf.pdf

[707] U.S. Energy Information Administration. Levelized generation costs (2023). https://www.eia.gov/outlooks/aeo/additional_docs.php This is the primary reference. Costs that are not included in this

reference are based on the other references using a ratio approach.

[708] National Renewable Energy Laboratory (September 2022): U.S. Solar photovoltaic system cost and energy storage benchmark: Q1 2022. https://www.nrel.gov/docs/fy22osti/83586.pdf

[709] National Energy Technology Laboratory report. Cost and performance baseline for fossil energy plants (2015). https://www.netl.doe.gov/energy-analysis/details?id=729

[710] U.S. EIA: Levelized cost and levelized avoided cost of new generation resources (2019). https://www.eia.gov/outlooks/archive/aeo19/pdf/electricity_generation.pdf

[711] Most of the data is from EIA. A full list of references is provided in the text.

[712] Most of the data is from EIA. A full list of references is provided in the text.

[713] Oakridge National Laboratory Report. Environmental quality and U.S. power sector. Air quality, water quality, land use and environmental justice. https://www.energy.gov/sites/prod/files/2017/01/f34/Environment%20Baseline%20Vol.%202--Environmental%20Quality%20and%20the%20U.S.%20Power%20Sector--Air%20Quality%2C%20Water%20Quality%2C%20Land%20Use%2C%20and%20Environmental%20Justice.pdf

[714] Lazard. Lazard's LCOE analysis. Version 15. https://www.lazard.com/perspective/levelized-cost-of-energy-levelized-cost-of-storage-and-levelized-cost-of-hydrogen/

[715] Oakridge National Laboratory Report. Environmental quality and U.S. power sector. Air quality, water quality, land use and environmental justice. https://www.energy.gov/sites/prod/files/2017/01/f34/Environment%20Baseline%20Vol.%202--Environmental%20Quality%20and%20the%20U.S.%20Power%20Sector--Air%20Quality%2C%20Water%20Quality%2C%20Land%20Use%2C%20and%20Environmental%20Justice.pdf

[716] IRENA report. Renewable generation costs in 2018. https://www.irena.org/-/media/Files/IRENA/Agency/Publication/2019/May/IRENA_Renewable-Power-Generations-Costs-in-2018.pdf

[717] IEA Hydropower special market report. https://iea.blob.core.windows.net/assets/4d2d4365-08c6-4171-9ea2-8549fabd1c8d/HydropowerSpecialMarketReport_corr.pdf

[718] U.S. National Academies. What you need to know about energy. Geothermal. http://needtoknow.nas.edu/energy/energy-sources/renewable-sources/geothermal/

[719] U.S. DOE. Geothermal basics. https://www.energy.gov/eere/geothermal/geothermal-basics

[720] IPCC report. Contribution of Working Group III to the Fifth Assessment Report of the Intergovernmental Panel on Climate Change. https://www.ipcc.ch/site/assets/uploads/2018/02/ipcc_wg3_ar5_annex-iii.pdf#page=7

[721] Our World in Data. Natural gas prices. https://ourworldindata.org/grapher/natural-gas-prices

[722] IEA. Electric vehicles. Total cost of ownership tool. https://www.iea.org/data-and-statistics/data-tools/electric-vehicles-total-cost-of-ownership-tool

[723] Prices must be compared for the same trim for a fair comparison. Kona Electric (SEL trim) is 55% higher than the Kona conventional (SEL trim). Kona Electric (Limited trim) is 44% higher than the Kona conventional (Limited trim). Prices are from the Hyundai website for the year 2023 models. https://www.hyundaiusa.com/us/en/vehicles/kona-electric

[724] IEA. Electric vehicles. Total cost of ownership tool. https://www.iea.org/data-and-statistics/data-tools/electric-vehicles-total-cost-of-ownership-tool

[725] Argonne National Laboratory Report. Comprehensive total cost of ownership calculation for vehicles with different size classes and power trains. https://publications.anl.gov/anlpubs/2021/05/167399.pdf

[726] Final report for the European Consumer Organization. Calculating the total cost of ownership for consumers. https://www.beuc.eu/publications/beuc-x-2021-

039_electric_cars_calculating_the_total_cost_of_ownership_for_consumers.pdf

[727] MIT Energy Initiative. Insights into Future Mobility. http://energy.mit.edu/wp-content/uploads/2019/11/Insights-into-Future-Mobility.pdf

[728] IEA data. Comparative life cycle GHG emissions from a midsize BEV and ICE vehicle. https://www.iea.org/data-and-statistics/charts/comparative-life-cycle-greenhouse-gas-emissions-of-a-mid-size-bev-and-ice-vehicle

[729] The 50% reduction is too optimistic based on recent studies. Prior studies used the average carbon intensity of the grid. Of more relevance is the carbon intensity of the marginal electricity being used by electric cars. The marginal electricity is typically provided by coal power or natural gas. So, the carbon intensity of the marginal electricity is much higher. This translates to a much lower benefit from electric cars. https://www.pnas.org/doi/abs/10.1073/pnas.2116632119

[730] MIT Energy Initiative. Insights into Future Mobility. http://energy.mit.edu/wp-content/uploads/2019/11/Insights-into-Future-Mobility.pdf

[731] U.S. Department of Energy. Developing infrastructure to charge plug -in electric vehicles. https://afdc.energy.gov/fuels/electricity_infrastructure.html#dc

[732] A typical BEV has a travel range of 250 miles. A conventional vehicle has a travel range of 400 miles. https://www.fueleconomy.gov/feg/Find.do?action=sbsSelect

[733] Argonne National Laboratory Report. Comprehensive total cost of ownership calculation for vehicles with different size classes and power trains. https://publications.anl.gov/anlpubs/2021/05/167399.pdf

[734] MIT Energy Initiative. Insights into Future Mobility. http://energy.mit.edu/wp-content/uploads/2019/11/Insights-into-Future-Mobility.pdf

[735] IEA data. Well to wheels greenhouse gas emissions for cars by power trains. https://www.iea.org/data-and-statistics/charts/well-to-wheels-greenhouse-gas-emissions-for-cars-by-powertrains

[736] Toyota Prius Eco has a 40% superior fuel efficiency than Toyota Corolla. Toyota RAV4 hybrid has a 30% higher efficiency than

Toyota RAV4.
https://www.fueleconomy.gov/feg/Find.do?action=sbsSelect

[737] Hybrid electric cars use conventional fueling stations. They have a longer travel range because of higher fuel efficiency. https://www.fueleconomy.gov/feg/Find.do?action=sbsSelect

[738] Cellulosic Ethanol. Status and Innovation. https://www.osti.gov/servlets/purl/1364156

[739] IEA Energy Technology Network. Oil Refineries. https://iea-etsap.org/E-TechDS/PDF/P04_Oil%20Ref_KV_Apr2014_GSOK.pdf

[740] IEA Bioenergy. Advanced Biofuels. Potential for Cost reduction. https://www.ieabioenergy.com/wp-content/uploads/2020/02/T41_CostReductionBiofuels-11_02_19-final.pdf

[741] U.S. EPA. Lifecycle Greenhouse Gas Results. https://www.epa.gov/fuels-registration-reporting-and-compliance-help/lifecycle-greenhouse-gas-results

[742] Toyota Website. 2023 MIRAI. https://www.toyota.com/mirai/

[743] Fuel cell vehicle costs are high (MIRAI). H_2 fuel costs are currently high as well. H_2 price is currently above $10 per gallon of gasoline equivalent. This translates to high lifetime costs.

[744] U.S. Department of Energy. Lifecycle greenhouse gas emissions from small sports utility vehicles. https://www.hydrogen.energy.gov/pdfs/21003-life-cycle-ghg-emissions-small-suvs.pdf

[745] U.S. Department of Energy. 5 things to know when fueling up your fuel cell electric vehicle. https://www.energy.gov/eere/articles/5-things-know-when-filling-your-fuel-cell-electric-vehicle

[746] Mass transit requires far fewer vehicles, lower fuel consumption and lower maintenance for the same passenger miles travelled. This translates to lower upfront and lifetime costs.

[747] IEA Energy system transport. Well to wheel ghg intensity of motorized transportation modes. https://www.iea.org/energy-system/transport/rail? This reference compares the greenhouse gas emissions of cars with mass transit options. The data shows that mass transit can reduce greenhouse gases more efficiently in most parts of the world.

[748] U.S. Department of Transportation. Public transportation's role in responding to climate change (2010). https://www.transit.dot.gov/sites/fta.dot.gov/files/docs/PublicTransportationsRoleInRespondingToClimateChange2010.pdf Mass transit vehicles at a 70% occupancy can reduce the greenhouse gases by 60% or more compared to personal conventional cars.

[749] IEA. The future of hydrogen. https://www.iea.org/reports/the-future-of-hydrogen

[750] The capacity factor is also low for the equipment because of the intermittent electricity from solar and wind power. It can only be used during the periods when the wind is blowing, and sun is shining.

[751] U.S. EIA. Hydrogen explained. https://www.eia.gov/energyexplained/hydrogen/

[752] U.S. Energy.gov. Hydrogen Program. Applications/Technology validation https://www.hydrogen.energy.gov/program-areas/enduse

[753] McKinsey & Company Special Report, January 2022. The net zero transition. What it would cost? What it could bring? https://www.mckinsey.com/business-functions/sustainability/our-insights/the-economic-transformation-what-would-change-in-the-net-zero-transition

[754] International Renewable Energy Agency. World Energy transitions outlook: 1.5°C pathway. https://www.irena.org/publications/2021/Jun/World-Energy-Transitions-Outlook

[755] McKinsey & Company Special Report, January 2022. The net zero transition. What it would cost? What it could bring? https://www.mckinsey.com/business-functions/sustainability/our-insights/the-economic-transformation-what-would-change-in-the-net-zero-transition

[756] The $3.5 trillion represents the amount of additional cost compared to the amount spent in 2021. The report was published in 2022.

[757] These numbers were provided by McKinsey & Company in the report for the year 2020.

[758] A few examples. The study estimates that about $2 trillion of assets will be stranded in the power sector alone. Early

retirements of a massive number of productive assets will lead to bankruptcies, and credit defaults. This has the potential to rock the global financial system. Also, including adequate redundancy in energy systems will result in enormous additional costs. Retraining costs will also be high. The study estimated that the energy transition will lead to a gain of 200 million jobs and loss of 185 million jobs. A massive retraining will be required to balance the work force.

[759] U.S. EIA. Electricity data. https://www.eia.gov/international/data/world/electricity/electricity-generation

[760] IEA Report. Net zero by 2050: A roadmap for the energy sector. https://www.iea.org/reports/net-zero-by-2050

[761] IRENA report. World Energy transitions outlook: 1.5°C pathway. https://www.irena.org/publications/2021/Jun/World-Energy-Transitions-Outlook

[762] Refer to the earlier discussions for more details.

[763] Resource intensity includes all the resources required. Examples of resources include labor, materials, energy, and land.

[764] Electric cars are 40 to 50% more expensive for the same trim. The upfront cost of heat pumps is two to four times higher. https://www.iea.org/reports/the-future-of-heat-pumps/executive-summary#

[765] IEA Updated 2023 report: Net zero by 2050. https://www.iea.org/reports/net-zero-roadmap-a-global-pathway-to-keep-the-15-0c-goal-in-reach https://www.iea.org/commentaries/reaching-net-zero-emissions-demands-faster-innovation-but-weve-already-come-a-long-way

[766] Many of the commercially available low carbon technologies are not yet competitive. The rapid market uptake of these technologies has mostly been driven by massive subsidies.

[767] IEA Report. Net zero by 2050. A roadmap for the energy sector. https://www.iea.org/reports/net-zero-by-2050

[768] IRENA report. World Energy transitions outlook 1.5°C pathway. https://www.irena.org/publications/2021/Jun/World-Energy-Transitions-Outlook

[769] European Environment Agency. Resource use and materials. https://www.eea.europa.eu/en/topics/in-depth/resource-use-and-materials

[770] German Environment Agency (Umwelt Bundesamt). Resource use and its consequences. https://www.umweltbundesamt.de/en/topics/waste-resources/resource-use-its-consequences#

[771] Our World in Data. Global direct energy consumption. https://ourworldindata.org/grapher/global-primary-energy?country=~OWID_WRL

[772] J. R. Fleming. Eos, Transactions American Geophysical Union. Volume 79. Page 405. Year 1998. https://agupubs.onlinelibrary.wiley.com/doi/abs/10.1029/98EO00310

[773] E. Hawkins, P.D. Jones. Quarterly Journal of the Royal Meteorological Society. Volume 139. Page 1961. Year 2013. https://rmets.onlinelibrary.wiley.com/doi/10.1002/qj.2178 https://centaur.reading.ac.uk/32981/1/hawkins_jones_2013.pdf

[774] Our World in Data. Global direct energy consumption. https://ourworldindata.org/grapher/global-primary-energy?country=~OWID_WRL

[775] International Energy Agency Report, 2021. Net zero by 2050: A roadmap for the energy sector. https://www.iea.org/reports/net-zero-by-2050

[776] IEA Report. The role of critical minerals on clean energy transitions. https://www.iea.org/reports/the-role-of-critical-minerals-in-clean-energy-transitions

[777] World Nuclear Association. Mineral requirement for electricity generation. https://www.world-nuclear.org/information-library/energy-and-the-environment/mineral-requirements-for-electricity-generation.aspx

[778] Oakridge National Laboratory Report. Environmental quality and U.S. power sector. Air quality, water quality, land use and environmental justice. https://www.energy.gov/sites/prod/files/2017/01/f34/Environment%20Baseline%20Vol.%202--Environmental%20Quality%20and%20the%20U.S.%20Power%20Sector--

Air%20Quality%2C%20Water%20Quality%2C%20Land%20Use%2C%20and%20Environmental%20Justice.pdf

[779] IEA Updated 2023 report: Net zero by 2050. https://www.iea.org/reports/net-zero-roadmap-a-global-pathway-to-keep-the-15-0c-goal-in-reach

[780] IEA Electric Vehicles. Tracking Report (2022). https://www.iea.org/reports/electric-vehicles There are 16.5 million electric cars out of total of 1.2 billion cars.

[781] IEA Updated 2023 report: Net zero by 2050. https://www.iea.org/reports/net-zero-roadmap-a-global-pathway-to-keep-the-15-0c-goal-in-reach

[782] Many of the commercially available low carbon technologies are not yet competitive. The market uptake of these technologies has mostly been driven by massive subsidies.

[783] Our World in Data. Global direct primary consumption. https://ourworldindata.org/grapher/global-primary-energy?country=~OWID_WRL

[784] IEA Updated 2023 report: Net zero by 2050. https://www.iea.org/reports/net-zero-roadmap-a-global-pathway-to-keep-the-15-0c-goal-in-reach

[785] International Energy Agency Report, 2021. Net zero by 2050: A roadmap for the energy sector. https://www.iea.org/reports/net-zero-by-2050

[786] IEA Report. Energy Technology Perspectives 2023. https://www.iea.org/reports/energy-technology-perspectives-2023#

[787] Other popular net zero proposals are similar to the IEA proposal. I focus on the IEA proposal here. But the discussion is generically applicable.

[788] U.S. EIA data. Primary energy. https://www.eia.gov/international/data/world/total-energy/total-energy-consumption?

[789] U.S. EIA: Primary energy. https://www.eia.gov/international/data/world/total-energy/total-energy-consumption?

[790] IEA Report. Net zero by 2050. A roadmap for the energy sector. Page 119. https://www.iea.org/reports/net-zero-by-2050

[791] IEA Report. Energy Technology Perspectives 2023. Page 283. https://www.iea.org/reports/energy-technology-perspectives-2023#

[792] Data Brief. A global inventory of electricity infrastructures from 1980 to 2017. Volume 38. Page 107351. Year 2021. https://www.ncbi.nlm.nih.gov/pmc/articles/PMC8441158/pdf/main.pdf

[793] U.S. EIA data. Electricity data. https://www.eia.gov/international/data/world/electricity/electricity-generation

[794] The data for current levels is from the year 2022 (i.e., the most recent available data). IEA data. Tracking clean energy progress 2023. https://www.iea.org/reports/tracking-clean-energy-progress-2023

[795] IEA Updated 2023 report: Net zero by 2050. https://www.iea.org/reports/net-zero-roadmap-a-global-pathway-to-keep-the-15-0c-goal-in-reach

[796] IEA data. Global hydrogen review 2022. https://www.iea.org/reports/global-hydrogen-review-2022/executive-summary

[797] IEA Report, 2021. The role of critical minerals on clean energy transitions. https://www.iea.org/reports/the-role-of-critical-minerals-in-clean-energy-transitions

[798] IEA provides data based on capacity (per MW). But the capacity factor or availability of wind and solar power to generate electricity is lower than natural gas and coal power. Also, the life of solar and wind power is lower than natural gas and coal power. So, a solar or wind power plant with the same capacity as a fossil fuel power plant produces much lower electricity over the life. A more relevant approach is to compare the materials required per unit electricity produced by the power plants over the lifetime. I have converted the IEA data from materials required per capacity to materials required per unit electricity generated over the lifetime. I have used average data from IEA reports for the capacity factors and lifetimes. I have used the average of natural gas and coal power to represent fossil fuel power. https://www.iea.org/data-and-statistics/charts/average-annual-capacity-factors-by-technology-2018

https://www.iea.org/reports/energy-technology-perspectives-2023

[799] USGS Lithium data. https://pubs.usgs.gov/periodicals/mcs2023/mcs2023-lithium.pdf

[800] USGS Nickle data. https://pubs.usgs.gov/periodicals/mcs2023/mcs2023-nickel.pdf

[801] USGS Cobalt data. https://pubs.usgs.gov/periodicals/mcs2023/mcs2023-cobalt.pdf

[802] IEA Report. Energy Technology Perspectives 2023. Pages 86,87,90,96. https://www.iea.org/reports/energy-technology-perspectives-2023#

[803] IEA Report. Energy Technology Perspectives 2023. Page 62. https://www.iea.org/reports/energy-technology-perspectives-2023#

[804] Oakridge National Laboratory Report. Environmental quality and U.S. power sector. Air quality, water quality, land use and environmental justice. https://www.energy.gov/sites/prod/files/2017/01/f34/Environment%20Baseline%20Vol.%202--Environmental%20Quality%20and%20the%20U.S.%20Power%20Sector--Air%20Quality%2C%20Water%20Quality%2C%20Land%20Use%2C%20and%20Environmental%20Justice.pdf

[805] Sabin Center for Climate Law Report. Opposition to renewable energy facilities in United States. May 2023 edition. https://scholarship.law.columbia.edu/sabin_climate_change/200/

[806] IEA Report. Energy Technology Perspectives 2023. Pages 71-75. https://www.iea.org/reports/energy-technology-perspectives-2023#

[807] IEA Updated 2023 report: Net zero by 2050. https://www.iea.org/reports/net-zero-roadmap-a-global-pathway-to-keep-the-15-0c-goal-in-reach IEA Commentary. https://www.iea.org/commentaries/reaching-net-zero-emissions-demands-faster-innovation-but-weve-already-come-a-long-way

[808] Also, many of the commercially available low carbon technologies are not yet competitive. The market uptake of these technologies has mostly been driven by massive subsidies.

[809] IEA Updated 2023 report: Net zero by 2050. https://www.iea.org/reports/net-zero-roadmap-a-global-pathway-to-keep-the-15-0c-goal-in-reach

[810] IEA Report. Energy Technology Perspectives 2023. Page 50. https://www.iea.org/reports/energy-technology-perspectives-2023#

[811] Greencar.com. 20 truths about GM EV1 electric car. The first mass produced modern electric car, EV1 from General Motors, was available for purchase in 1996. Note, electric cars were first introduced more than a century ago. https://web.archive.org/web/20090123001021/http://www.greencar.com/features/gm-ev1/

[812] The time required for major innovations cannot be predicted. The new technology will need to be available at a comparable cost to the incumbent. Also, it will need to provide a similar convenience to the user.

[813] IEA Report, 2021. The role of critical minerals on clean energy transitions. https://www.iea.org/reports/the-role-of-critical-minerals-in-clean-energy-transitions

[814] IEA Report. Energy Technology Perspectives 2023. Pages 86,87,90,96. https://www.iea.org/reports/energy-technology-perspectives-2023#

[815] IEA Report. Energy Technology Perspectives 2023. Pages 86,87,90,96. https://www.iea.org/reports/energy-technology-perspectives-2023#

[816] Fossil fuels technologies have massive help from nature. The popular low carbon technologies do not. So, they have a markedly higher resource intensity.

[817] IEA report (2022). Accelerating sector transitions through stronger international collaborations. https://www.iea.org/reports/breakthrough-agenda-report-2022

[818] United Nations. Covid vaccines. Widening inequality and millions vulnerable. https://news.un.org/en/story/2021/09/1100192

[819] Our World in data. Coronavirus vaccinations. https://ourworldindata.org/covid-vaccinations

[820] Public trust can be discussed in terms of the percent of the population that has high or moderately high trust in the government.

[821] OECD data. Trust in government. https://data.oecd.org/gga/trust-in-government.htm

[822] OECD Trust Survey Report. https://www.oecd.org/governance/trust-in-government/

[823] International Monetary Fund. General Government Debt. https://www.imf.org/external/datamapper/GG_DEBT_GDP@GDD/

[824] These are based on the crucial facts about the science and practical aspects of climate and energy.

[825] Details are discussed in a previous chapter.

[826] IEA Energy system transport. Well to wheel ghg intensity of motorized transportation modes. https://www.iea.org/energy-system/transport/rail? This reference compares the greenhouse gas emissions of cars with mass transit options. The data shows that mass transit can reduce greenhouse gases more efficiently in most parts of the world.

[827] IEA report. The future of heat pumps. https://www.iea.org/reports/the-future-of-heat-pumps/executive-summary

[828] IEA Commentary. https://www.iea.org/commentaries/reaching-net-zero-emissions-demands-faster-innovation-but-weve-already-come-a-long-way

[829] The total budget for the less cost-effective options should be ten percent or less of the total allocated for the energy transition.

[830] IEA: Comparative life cycle GHG emissions from a midsize BEV and ICE vehicle (2021). https://www.iea.org/data-and-statistics/charts/comparative-life-cycle-greenhouse-gas-emissions-of-a-mid-size-bev-and-ice-vehicle

[831] U.S. EPA. Climate adaptation plans. https://www.epa.gov/climate-adaptation/climate-adaptation-plans Strategies for climate change adaptation. https://www.epa.gov/arc-x/strategies-climate-change-adaptation

[832] United Nations Environment programme. Five ways countries can adapt to climate crisis. https://www.unep.org/news-and-stories/story/5-ways-countries-can-adapt-climate-crisis

[833] Refer to previous chapter for details about the net zero by 2050 proposal.

[834] IEA states that if the global society was to achieve net zero goals by 2050, there is no need for any new oil and gas projects. For this scenario to be true, the global society will need to meet all the extraordinary challenges.

[835] It will be critical to focus on cost-effective options and an efficient use of resources.

[836] They have a markedly higher resource intensity than fossil fuel technologies. Refer to previous chapter for details.

[837] This is only possible when multiple options are competitive and can satisfy all the required conditions for a sustainable energy transition. Refer to the list of cost-effective options in the power sector. Multiple options are available in this crucial sector. Multiple options will also very likely be available in other sectors with time (technology advances).

[838] Diverse options require diverse resources. For example, solar power requires a different set of critical minerals compared to wind power. The excessive use of any resource will cause a large impact. The use of different resources dilutes the impact from each individual resource.

[839] A far lower (less than half) amount of greenhouse gases would have been released.

[840] Only options that also satisfy other criteria should be used. For example, energy efficient options should only be encouraged if they are cost-effective and show high overall efficiency (i.e., the cart should not be placed before the horse).

[841] IEA Energy system transport. Well to wheel ghg intensity of motorized transportation modes. https://www.iea.org/energy-system/transport/rail? This reference compares the greenhouse gas emissions of cars with mass transit options. The data shows that mass transit can reduce greenhouse gases more efficiently in most parts of the world.

[842] United Nations Environment Programme. Why the global fight to tackle food waste has just begun (Sept. 2022). https://www.unep.org/news-and-stories/story/why-global-fight-tackle-food-waste-has-only-just-begun#

[843] United States Agency for International Development (USAID). Clean energy solutions for agriculture productivity. https://www.usaid.gov/energy/powering-agriculture/opportunity

[844] More than 190 countries have ratified the Paris Climate Accord. United Nations Climate Change (UNFCC): Paris Agreement – Status of Ratification. https://unfccc.int/process/the-paris-agreement/status-of-ratification

[845] It is important to distinguish between early adoption and wide deployment. Governments should provide incentives for early adoption, which is key to lower the prices for new technologies. But the government should not prioritize high-cost technologies for a wide scale deployment.

[846] European Commission. Energy. Solar Energy. https://energy.ec.europa.eu/topics/renewable-energy/solar-energy_en

[847] Whitehouse.gov. Building a clean energy economy. https://www.whitehouse.gov/wp-content/uploads/2022/12/Inflation-Reduction-Act-Guidebook.pdf

[848] U.S. EIA report. Cost and performance characteristics of new generation resources in the annual energy outlook 2022. https://www.eia.gov/outlooks/aeo/assumptions/pdf/table_8.2.pdf

[849] U.S. EIA report. Levelized cost of new generation resources in the annual energy outlook 2022. https://www.eia.gov/outlooks/aeo/pdf/electricity_generation.pdf

[850] IEA report. Projected costs of generating electricity. https://iea.blob.core.windows.net/assets/ae17da3d-e8a5-4163-a3ec-2e6fb0b5677d/Projected-Costs-of-Generating-Electricity-2020.pdf

[851] Decades of data shows an excellent safety profile for nuclear power when compared to the other options. See the earlier discussion on nuclear power.

[852] IEA Key statistics. Germany. https://www.iea.org/countries/germany

[853] Why? Because dispatchable sources are critical to maintain a high grid reliability. Coal and nuclear power are dispatchable sources. While solar and wind power are intermittent sources.

[854] EMBER electricity data explorer. https://ember-climate.org/data/data-tools/data-explorer/

[855] The intensity for the world was 437 gCO_2/kWh in 2022. For Germany it was only a little lower at 385 gCO_2/kWh.

[856] EMBER electricity data explorer. https://ember-climate.org/data/data-tools/data-explorer/

[857] California energy commission. Total system electricity generation spreadsheet. https://www.energy.ca.gov/data-reports/energy-almanac/california-electricity-data/2021-total-system-electric-generation

[858] Why? Because California needs a large use of natural gas power to maintain grid reliability. Also, the efficiency decreases and emissions increase when natural gas plants are operated sub-optimally to stabilize intermittent power generation.

[859] California air resources board. California greenhouse gas emissions for 2000 to 2020. https://ww2.arb.ca.gov/sites/default/files/classic/cc/inventory/2000-2020_ghg_inventory_trends.pdf

[860] California energy commission. Total system electricity generation spreadsheet. https://www.energy.ca.gov/data-reports/energy-almanac/california-electricity-data/2021-total-system-electric-generation

[861] California air resources board. California greenhouse gas emissions for 2000 to 2020. https://ww2.arb.ca.gov/sites/default/files/classic/cc/inventory/2000-2020_ghg_inventory_trends.pdf

[862] A dilution of the energy mix is key to reduce future impacts from the environment. This will require a diverse use of options.

[863] Electric cars have 40% to 50% higher cost than conventional cars. See the earlier discussion on electric cars. The upfront cost of heat pumps is 2 to 4 times more than the conventional option. IEA report. The future of heat pumps. https://www.iea.org/reports/the-future-of-heat-pumps/executive-summary#

[864] EMBER electricity data explorer. https://ember-climate.org/data/data-tools/data-explorer/ A medium carbon intensity is between 250 to 500 gCO_2/kWh. High intensity is above 500 gCO_2/kWh.

[865] IEA smart grids. https://www.iea.org/energy-system/electricity/smart-grids#

[866] U.S. Senate hearing. https://www.energy.senate.gov/services/files/1D618EDD-7CED-4BC5-8F09-C8F0668FE608

[867] Joint comments of four major U.S. electricity system operators to the EPA. https://www.pjm.com/-/media/documents/other-fed-state/20230808-comments-of-joint-isos-rtos-docket-epa-hq-oar-2023-0072.ashx

[868] International Energy Agency Report, 2021. Net zero by 2050: A roadmap for the energy sector. https://www.iea.org/reports/net-zero-by-2050 The electrical grids will need to more than double for a net zero world.

[869] North American Electric Reliability Corporation. 2023 ERO reliabilities priority report. https://www.nerc.com/comm/RISC/Related%20Files%20DL/RISC_ERO_Priorities_Report_2023_Board_Approved_Aug_17_2023.pdf

[870] International Energy Agency Report, 2021. Net zero by 2050: A roadmap for the energy sector. https://www.iea.org/reports/net-zero-by-2050

[871] Environmental and Energy science institute. Comparing U.S and Chinese electric vehicle policies. https://www.eesi.org/articles/view/comparing-u.s.-and-chinese-electric-vehicle-policies

[872] EMBER electricity data explorer. https://ember-climate.org/data/data-tools/data-explorer/ Medium intensity is between 250 to 500 gCO_2/kWh. High intensity is above 500 gCO_2/kWh.

[873] IEA data. Comparative life cycle GHG emissions from a midsize BEV and ICE vehicle. https://www.iea.org/data-and-statistics/charts/comparative-life-cycle-greenhouse-gas-emissions-of-a-mid-size-bev-and-ice-vehicle

[874] The reduction with electric cars is 50% for a grid with the global average intensity (~450 gCO_2/kWh). It is 7% for a grid with an intensity of 800 gCO_2/kWh.

[875] The 50% reduction reported by IEA is likely too optimistic based on recent studies. Prior studies used the average carbon intensity of the grid. Of more relevance is the carbon intensity of the marginal electricity being used by electric cars. The marginal

electricity is typically provided by coal power or natural gas. So, the carbon intensity of the marginal electricity is much higher. This translates to a much lower benefit from electric cars. https://www.pnas.org/doi/abs/10.1073/pnas.2116632119

[876] Details are provided in an earlier chapter.

[877] IEA Energy system transport. Well to wheel ghg intensity of motorized transportation modes. https://www.iea.org/energy-system/transport/rail? This reference compares the greenhouse gas emissions of cars with mass transit options. The data shows that mass transit can reduce greenhouse gases more efficiently in most parts of the world.

[878] China Briefing (June 2023). https://www.china-briefing.com/news/china-considers-extending-its-ev-subsidies-to-2023/

[879] Reuters (September 2023). EU to investigate flood of Chinese electric cars, weigh tariffs. https://www.reuters.com/world/europe/eu-launches-anti-subsidy-investigation-into-chinese-electric-vehicles-2023-09-13/

[880] IEA Key energy statistics China. https://www.iea.org/countries/China

[881] Our World in data. Electricity production by source. China. https://ourworldindata.org/energy/country/china#what-sources-does-the-country-get-its-electricity-from

[882] IEA Report. Coal 2022. Figure: Global Coal consumption. https://www.iea.org/reports/coal-2022/executive-summary

[883] CREA Global Energy monitor. China permits 2 new coal power plants per week in 2022. https://energyandcleanair.org/wp/wp-content/uploads/2023/02/CREA_GEM_China-permits-two-new-coal-power-plants-per-week-in-2022.pdf

[884] China has also been shutting down many old less efficient plants. But overall, the electricity generation from coal has increased massively since the last decade or so. See previous IEA and Our World in Data references.

[885] Coal power alone emits more than 20% of the global greenhouse gas emissions. https://www.iea.org/reports/global-energy-review-CO_2-emissions-in-2021-2 Coal power emits 10.5 billion tons and total coal use results in 15.3 billion tons of ghg emissions.

[886] Ember electricity data explorer. https://ember-climate.org/data/data-tools/data-explorer/ China's carbon intensity of electricity is 20% higher than the world average.

[887] Reports vary by a lot about the ghg reduction that can be achieved by electric cars. Some recent reports suggest a reduction of less than 10%. Previous studies were based on average carbon intensity of grid. This provides a benefit that is too optimistic. Using the carbon intensity of the marginal electricity used by electric cars is more relevant. State of the art is to focus on using carbon intensity of the marginal electricity production. Such studies indicates that electric cars reduce emissions by only about 10% in China.

[888] IScience. Revisiting electric vehicle life cycle greenhouse gas emissions in China. A marginal emission perspective. Volume 26. Page 106565. Year 2023. https://www.sciencedirect.com/science/article/pii/S2589004223006429

[889] Sustainability. Cradle to grave life cycle analysis of ghg emissions of light duty passenger vehicles in China. Volume 15. Page 2627. Year 2023. https://www.mdpi.com/2071-1050/15/3/2627

[890] The International council on clean transportation (ICCT). A global comparison of the lifecycle gas emissions of combustion engine and electric passenger cars. https://theicct.org/sites/default/files/publications/Global-LCA-passenger-cars-jul2021_0.pdf This reference uses average carbon intensity of electrical grid and provides very optimistic values.

[891] China Briefing. https://www.china-briefing.com/news/china-considers-extending-its-ev-subsidies-to-2023/

[892] China has about 10.7 million electric cars on the road currently. The subsidies have ranged from about 4000$ to $9000 for the cars. So, the subsidies for electric cars have been more than 43 billion dollars.

[893] Electric cars can reduce between 5% to 40% CO_2 emissions compared to conventional vehicles in China. Nuclear power can reduce 95% emissions when replacing coal power. China has about 10.7 million cars on the road. It has spent over $43 billion in subsidies for those electric cars. 10.7 million electric cars can

reduce 13 MM tons of CO_2 per year. This uses the extremely optimistic CO_2 reduction of 40%. Replacing just 2 GW of coal power capacity with nuclear power can reduce the same amount of CO_2 per year. The net upfront cost for this is about $9 billion in China. This cost is several times lower than the subsidies for electric cars. This also applies to replacing coal power with solar, wind, hydropower, or natural gas power. For the same initial investment, at-least five times more CO_2 can be reduced by a focus on the power sector instead of electric cars. https://www.yicaiglobal.com/news/china-approves-two-nuclear-power-projects-at-a-cost-of-usd115-billion Also, an expansion of mass transit and hybrid cars are far superior to reduce emissions from the transportation sector.

[894] IEA Energy system transport. Well to wheel ghg intensity of motorized transportation modes. https://www.iea.org/energy-system/transport/rail? This reference compares the greenhouse gas emissions of cars with mass transit options. The data shows that mass transit can reduce greenhouse gases more efficiently in most parts of the world. This is especially true in China.

[895] Facts and Details. Road congestion and traffic jams in China. https://factsanddetails.com/china/cat13/sub86/entry-8368.html

[896] Wikipedia. List of countries by vehicles per capita. https://en.wikipedia.org/wiki/List_of_countries_by_vehicles_per_capita

[897] Recall, the local impact from climate change does not depend on where the CO_2 is released or eliminated. It depends on the net amount of global CO_2 emissions and specifics of the location.

[898] IEA report (2023). World energy investment. https://www.iea.org/reports/world-energy-investment-2023/overview-and-key-findings

[899] The investment in upgrading the grids is still very low when considering the two bills combined. The American Society of Civil Engineers has estimated that the grid investment trends observed in 2021 would lead to a funding gap of $200 billion in 2029. https://www.energy.gov/gdo/bipartisan-infrastructure-law#:~:text=BIL%20Provision%2040103(b)%20%2D,storage%2C%20and%20distribution%20infrastructure%20to

https://infrastructurereportcard.org/wp-content/uploads/2020/12/Energy-2021.pdf

[900] The White House Fact Sheet. October 30, 2023. https://www.whitehouse.gov/briefing-room/statements-releases/2023/10/30/fact-sheetbiden-harris-administration-announces-historic-investment-to-bolster-nations-electric-grid-infrastructure-cut-energy-costs-for-families-and-create-good-paying-jobs/

[901] U.S. Department of Energy. Inflation reduction act. Grid deployment office. https://www.energy.gov/gdo/inflation-reduction-act#:~:text=Through%20the%20Inflation%20Reduction%20Act,transmission%20lines%20across%20the%20country.

[902] The last three years include 2021,2022, 2023. https://www.iea.org/reports/world-energy-investment-2023 Overview and Key Findings.

[903] United Nations report. A world of debt. https://unctad.org/publication/world-of-debt

[904] World economic forum. What is global debt and how high is it now. https://www.weforum.org/agenda/2023/10/what-is-global-debt-why-high/

[905] I will discuss the policies related to the large subsidies for electric cars as an example. A reliable low carbon electrical grid and mass transit deserve the highest priority and resource allocation. But that is not happening. Instead, critical resources are being pulled away for electric cars in many regions. This is not a cost effective or efficient approach to address climate change.

[906] The impact from poor policies will be less noticeable in the short term. This is because fossil fuels will dominate the energy mix in the short term. So, fossil fuels will determine the energy availability and cost in the short term. The impact from poor policies will increase as the world continues to shift to low carbon energy.

[907] In this section, my focus is on global policies related to the energy transition. Most countries have already agreed to the energy transition. So, the misinformation from skeptics is not relevant. The world has moved on from a practical viewpoint. The debate is about the best path to address climate change.

[908] IPCC AR6. Climate Change 2021. The Physical Science Basis. https://www.ipcc.ch/report/ar6/wg1/downloads/report/IPCC_AR6_WGI_SPM_final.pdf

[909] Net zero by 2050. Details with references are available in an earlier chapter.

[910] IEA Updated 2023 report: Net zero by 2050. https://www.iea.org/reports/net-zero-roadmap-a-global-pathway-to-keep-the-15-0c-goal-in-reach IEA commentary. https://www.iea.org/commentaries/reaching-net-zero-emissions-demands-faster-innovation-but-weve-already-come-a-long-way

[911] Many of the commercially available low carbon technologies are not yet competitive. The rapid market uptake of these technologies has mostly been driven by massive subsidies.

[912] Mass transit is cost-effective and decreases the use of energy and resources compared to personal cars. Policies should focus on the use of shuttle vans. This will ensure an excellent area coverage and low wait-times. Shuttle vans do not require any special infrastructure. This option can be quickly deployed.

[913] Hybrid cars have a far lower upfront cost. They also offer far more convenience. So, hybrid cars are likely to be adopted at a much faster pace by the public.

[914] The carbon intensity in the U.S. was 367 gCO_2/kWh in 2022. Global average of the world was 437 gCO_2/kWh. https://ember-climate.org/data/data-tools/data-explorer/

[915] Low carbon intensity means a value below 150 gCO_2/kWh.

[916] U.S. EIA FAQs. How much of the U.S. CO_2 emissions are associated with electricity generation? https://www.eia.gov/tools/faqs/faq.php?id=77&t=11 Data is for the year 2022.

[917] U.S. EIA FAQs. What is the U.S. electricity generation by energy source? https://www.eia.gov/tools/faqs/faq.php?id=427&t=3 Data is for the year 2022.

[918] PJM report (2023). Energy transition in PJM. https://www.pjm.com/-/media/library/reports-notices/special-reports/2023/energy-transition-in-pjm-resource-retirements-replacements-and-risks.ashx

[919] Historical data is informative. It shows that there are several two+ consecutive days in which the combined electricity from solar

and wind power is dramatically reduced. This is true even when considering an extended combined grid for the United States.

[920] North American Electric Reliability Corporation. 2023 ERO reliabilities priority report. https://www.nerc.com/comm/RISC/Related%20Files%20DL/RISC_ERO_Priorities_Report_2023_Board_Approved_Aug_17_2023.pdf

[921] Our World in data. Natural gas prices. https://ourworldindata.org/grapher/natural-gas-prices

[922] U.S. EIA Annual energy outlook. Additional AEO documentation. https://www.eia.gov/outlooks/aeo/additional_docs.php

[923] United Nations Economic Commission for Europe (March 2022). Lifecycle assessments of electricity generation options. https://unece.org/sed/documents/2021/10/reports/life-cycle-assessment-electricity-generation-options

[924] Efforts are underway to decrease the cost and improve the performance of CCS. https://www.netl.doe.gov/carbon-management/carbon-storage/faqs/carbon-storage-faqs

[925] Electric cars are 40 to50% more expensive for the same trim. The upfront cost of heat pumps is two to four times higher. https://www.iea.org/reports/the-future-of-heat-pumps/executive-summary#

[926] Electric cars can reduce emissions by less than 50% for a grid with medium carbon intensity. The reduction can be greater than 80% for grids with a low carbon intensity.

[927] IEA data. Comparative life cycle GHG emissions from a midsize BEV and ICE vehicle. https://www.iea.org/data-and-statistics/charts/comparative-life-cycle-greenhouse-gas-emissions-of-a-mid-size-bev-and-ice-vehicle

[928] PJM report (2023). Energy transition in PJM. https://www.pjm.com/-/media/library/reports-notices/special-reports/2023/energy-transition-in-pjm-resource-retirements-replacements-and-risks.ashx

[929] This can be a major concern when the grid is most vulnerable. Examples are severe weather events and periods during which weather markedly lowers the output from renewables.

[930] U.S. Department of energy. https://www.energy.gov/eere/vehicles/articles/fotw-1208-oct-18-2021-life-cycle-greenhouse-gas-emissions-2020-electric https://www.hydrogen.energy.gov/pdfs/21003-life-cycle-ghg-emissions-small-suvs.pdf

[931] Environmental Research Letters. The role of pick-up truck electrification in the decarbonization of light duty vehicles. https://iopscience.iop.org/article/10.1088/1748-9326/ac5142 Data is from Figure S9b. Use of marginal carbon intensity in the estimate indicates a reduction of 41% for electric cars. Use of average emissions gives an estimate of 65% reduction. The authors admit that the use of marginal carbon intensity is more practical.

[932] IEA Energy system transport. Well to wheel ghg intensity of motorized transportation modes. https://www.iea.org/energy-system/transport/rail?

[933] Well to wheel does not include manufacture of vehicles, and maintenance. But the contribution from these components is less than 10% for buses and minibuses. The contribution is higher for cars. https://theicct.org/wp-content/uploads/2023/02/lca-ghg-fs2-emissions-hdv-fuels-europe-feb23.pdf https://www.sciencedirect.com/science/article/pii/S1361920919302792

[934] U.S. Department of transportation. Expand public transportation systems and offer incentives. https://www.transportation.gov/mission/health/Expand-Public-Transportation-Systems-and-Offer-Incentives

[935] U.S. households have an average of 2 cars. https://www.bts.gov/archive/publications/highlights_of_the_2001_national_household_travel_survey/section_01# Use of mass transit and ride sharing can reduce the need for an extra car in many households.

[936] A high occupancy can be ensured by policies that greatly incentivize the use of mass transport.

[937] See previous discussion for details and references.

[938] Range anxiety is not an issue with hybrid cars. The wait times for hybrid cars for refueling are several times lower than the recharging of electric cars.

[939] Recall, my use of the term "hybrid cars" in this book includes only those cars that do not require external charging. So, no plug-in hybrids.

[940] The much faster embrace of hybrid cars will lead to faster greenhouse gas emission reductions. The retail price of electric cars is expected to drop markedly in the next several years. Also, the convenience is expected to improve. The grid will be more reliable and have much lower carbon intensity. This will greatly increase the priority ranking of electric cars in the next phase.

[941] Our World in data. Energy use per person, 2022. https://ourworldindata.org/grapher/per-capita-energy-use?tab=table

[942] Materials, and goods require energy. For example, energy is required to extract, process and manufacture. So, a lower need for materials and goods translates to a lower energy use.

[943] Environmental Research Letters (2021). U.S. residential heat pumps. The private economic potential and its emissions, health, and grid impacts. Vol. 16, 084026. https://iopscience.iop.org/article/10.1088/1748-9326/ac10dc/pdf

[944] United States has a medium carbon intensity when considering all the states. But there is a lot of diversity among the states. Many states have a low carbon intensity grid.

[945] EIA. Today in energy. Carbon intensity of U.S. power generation continues to fall but varies widely by state. https://www.eia.gov/todayinenergy/detail.php?id=53819

[946] If electrification is carefully planned, there is a much smaller risk of grid issues. All electrification options are not alike. So, policies should also be different. Superior options to reduce greenhouse gases are available as an alternative to electric cars. That is not the case for heat pumps.

[947] Details are available in an earlier chapter.

[948] It is easier to satisfy energy needs when the use is lower.

[949] U.S. EIA report. International energy outlook 2023. https://www.eia.gov/outlooks/ieo/

[950] The practical scenarios of IEA show a large role of oil and gas in the foreseeable future. IEA world energy outlook 2023. https://www.iea.org/reports/world-energy-outlook-2023 IEA has estimated that no new oil and gas fields are necessary in its net

zero scenario. But that scenario requires the global society to magically address many extraordinary challenges. There is no credible data to suggest that we can do so. It will require several miracles. I have discussed details earlier.

[951] There should be no government subsidies for such options.

[952] The new technologies cannot be too inferior. I will discuss two examples. The upfront cost or lifetime cost should not be more than 15% higher for high-cost options such as cars. The lifetime cost should not be more than 25% higher for electricity. The performance should be comparable. This means that the new technology should provide a similar level of convenience.

[953] National Archives. Founders online. https://founders.archives.gov/documents/Washington/05-16-02-0376

[954] Poor-quality information is typically used to support a specific narrative.

[955] Refer to the previous chapter for details.

[956] The Lancet: Planetary health. Volume 5, Issue 7, 415-425, 2021. https://www.thelancet.com/journals/lanplh/article/PIIS2542-5196(21)00081-4/fulltext

[957] This study is consistent with the fact that climate change is causing the cold extremes to decrease. IPCC AR6. Climate Change 2021. The Physical Science Basis. https://www.ipcc.ch/report/ar6/wg1/downloads/report/IPCC_AR6_WGI_SPM_final.pdf

[958] They use the assumption that the global uptake of low carbon energy will be very rapid. Based on this assumption they conclude that the use of fossil fuels will rapidly drop to very low levels. This, in turns, leads to the conclusion that now new oil and gas fields will be required.

[959] Refer to the previous chapter for details.

[960] Refer to the previous chapter for details.

[961] Refer to the previous chapter for details.

[962] Hydropower and geothermal power have far fewer challenges compared to solar and wind power to provide 24X7 electricity, even when used at very high levels.

[963] Basically, the neighboring country grids are used as energy storage.

[964] OECD Report. Comparing nuclear accident risks with those from other energy sources. https://www.oecd-nea.org/jcms/pl_14538/comparing-nuclear-accident-risks-with-those-from-other-energy-sources?details=true

[965] U.S. EIA: Electricity data. https://www.eia.gov/international/data/world/electricity/electricity-generation Average share of nuclear power from 1980 to 2021 was 14%.

[966] What is the recent status? More and more countries are committing to either extend the life of existing nuclear power plants or build new ones. IEA Report. Nuclear power. https://www.iea.org/energy-system/electricity/nuclear-power#

[967] Major projects have a budget of more than hundred million dollars. They require very large resources over a short timeframe. This is why major projects are very complex.

[968] The energy industry has spent hundreds of billions of dollars on developing energy technologies and deploying them over the decades. This is unique to the energy industry. Thus, the energy industry provides a unique access to the relevant expertise. So, most energy experts have either a current or past affiliation with the energy industry.

[969] Developing energy expertise is a slow process. It takes a lot of time to gain access to the required experience. It takes more than ten years for an energy technology to move from the idea phase to commercial application. Developing expertise in technology evaluation requires at-least five years. Major projects take more than five years from initiation to completion.

[970] Google Scholar. https://scholar.google.com This database can be used to access the abstracts of most papers.

[971] Nature. Climate simulations recognize the hot model problem. https://www.nature.com/articles/d41586-022-01192-2

[972] Science. Use of too hot climate models exaggerates impact of global warming. https://www.science.org/content/article/use-too-hot-climate-models-exaggerates-impacts-global-warming

[973] IPCC Report: Climate Change 2021, Physical Basis. Contribution of working group I to the sixth assessment report. https://www.ipcc.ch/report/ar6/wg1/#FullReport

[974] History informs us that sensational claims are rarely accurate. The scientific community can fairly quickly reproduce the work on the rare occasions that they are accurate. If the sensational claims are found to be accurate, they quickly become the expert consensus. So, there is very little downside to ignore sensational claims from isolated papers.

[975] For example, nobody could predict the COVID-19 pandemic.

[976] There are several extraordinary challenges. A rapid access to an unprecedented number of resources. An unprecedented level of success in technology innovation. Ensuring robust balance between energy supply and demand. The papers do not include the massive impacts from these challenges. They tacitly assume success. See previous chapters for details.

[977] I have provided details in an earlier chapter.

[978] I have provided details in an earlier chapter.

[979] I have discussed earlier how the global data on disasters does not support this.

[980] The Lancet, Planetary health. Climate anxiety in children and young people and their beliefs about governments response to climate change. A survey. Volume 5. Page E 863. Year 2021. https://www.thelancet.com/journals/lanplh/article/PIIS2542-5196(21)00278-3/fulltext

[981] External publications are not a focus in the energy industry.

[982] Relevant expertise includes many years of experience with the following. 1) Evaluating energy technologies. 2) Developing and deploying energy technologies. 3) Major projects. Such experience can only be gained after working for many years in the industry. Academics do not have the opportunity to gain the relevant expertise.

[983] IEA report (September 2022). World energy employment. https://www.iea.org/reports/world-energy-employment/executive-summary

[984] IEA report (2023). World energy investment. https://www.iea.org/reports/world-energy-investment-2023/overview-and-key-findings

[985] The scale up of the low carbon energy solutions at the proposed speed is very unlikely because of the unprecedented challenges.

The halt of new fossil fuel production is based on the faulty premise that clean energy will scale up very rapidly. The halt of new fossil fuel production will be premature because of the very likely delay. This will cause a shortage of fossil fuels. The resulting disruption of energy supply will increase the costs.

[986] A major fraction of the global population belongs to these categories.

[987] Energy prices impact the entire economy. A rise in energy costs also leads to a rise in costs of goods. This is because energy is required to produce and transport goods. The poor and the middle class are most affected by such cost increases.

[988] Why? Because scientists and institutions have not pushed back against the false narratives. Two false narratives are of concern. 1) A rapid energy transition will not be too difficult. 2) It will help the poor and middle class by decreasing the costs to the society.

[989] The extreme views of the activists and skeptics are causing the misinformation crisis. The poor policies resulting from the misinformation will be a major hurdle to resolving our climate and energy problems. It is crucial that that we address these problems robustly.

[990] The use of speculative assumptions is very unlikely to provide any meaningful information.

Printed in Great Britain
by Amazon